# Conference Proceedings of the Society for Experimental Mechanics Series

*Series Editor*

Kristin B. Zimmerman, Ph.D.
Society for Experimental Mechanics, Inc.
Bethel, CT, USA

More information about this series at http://www.springer.com/series/8922

Alan T. Zehnder • Jay Carroll • Kavan Hazeli • Ryan B. Berke
Garrett Pataky • Matthew Cavalli • Alison M. Beese • Shuman Xia
Editors

# Fracture, Fatigue, Failure and Damage Evolution, Volume 8

## Proceedings of the 2016 Annual Conference on Experimental and Applied Mechanics

Springer

*Editors*
Alan T. Zehnder
Cornell University
Ithaca, NY, USA

Jay Carroll
Sandia National Laboratories
Albuquerque, NM, USA

Kavan Hazeli
John Hopkins University
Baltimore, MD, USA

Ryan B. Berke
School of Engineering
University of Liverpool
Liverpool, UK

Garrett Pataky
Department of Mechanical Engineering
Clemson University
Clemson, SC, USA

Matthew Cavalli
Department of Mechanical Engineering
University of North Dakota
Grand Forks, ND, USA

Alison M. Beese
Pennsylvania State University
State College, PA, USA

Shuman Xia
Georgia Institute of Technology
Atlanda, GA, USA

ISSN 2191-5644 ISSN 2191-5652 (electronic)
Conference Proceedings of the Society for Experimental Mechanics Series
ISBN 978-3-319-42194-0 ISBN 978-3-319-42195-7 (eBook)
DOI 10.1007/978-3-319-42195-7

Library of Congress Control Number: 2016949027



Printed on acid-free paper

This Springer imprint is published by Springer Nature
The registered company is Springer International Publishing AG Switzerland

# Preface

*Fracture, Fatigue, Failure and Damage Evolution* represents one of ten volumes of technical papers presented at the SEM 2016 Annual Conference & Exposition on Experimental and Applied Mechanics organized by the Society for Experimental Mechanics and held in Orlando, FL, June 6–9, 2016. The complete proceedings also includes volumes on *Dynamic Behavior of Materials*; *Challenges in Mechanics of Time-Dependent Materials*; *Advancement of Optical Methods in Experimental Mechanics*; *Experimental and Applied Mechanics*; *Micro-and Nanomechanics*; *Mechanics of Biological Systems and Materials*; *Mechanics of Composite and Multifunctional Materials*; *Residual Stress, Thermomechanics and Infrared Imaging, Hybrid Techniques and Inverse Problems*; and *Joining Technologies for Composites and Dissimilar Materials*.

Each collection presents early findings from experimental and computational investigations on an important area within experimental mechanics, fracture and fatigue being one of these areas.

Fatigue and fracture are two of the most critical considerations in engineering design. Understanding and characterizing fatigue and fracture has remained as one of the primary focus areas of experimental mechanics for several decades. Advances in experimental techniques, such as digital image correlation, acoustic emissions, and electron microscopy, have allowed for deeper study of phenomena related to fatigue and fracture. This volume contains the results of investigations of several aspects of fatigue and fracture such as microstructural effects, the behavior of interfaces, the behavior of different and/or complex materials such as composites, and environmental and loading effects. The collection of experimental mechanics research included here represents another step toward solving the long-term challenges associated with fatigue and fracture.


| | |
|---|---|
| Ithaca, NY, USA | Alan T. Zehnder |
| Albuquerque, NM, USA | Jay Carroll |
| Baltimore, MD, USA | Kavan Hazeli |
| Liverpool, UK | Ryan B. Berke |
| Clemson, SC, USA | Garrett Pataky |
| Grand Forks, ND, USA | Matthew Cavalli |
| State College, PA, USA | Allison M. Beese |
| Atlanta, GA, USA | Shuman Xia |

# Contents

# Chapter 1
# Fatigue Damage Precursor Identification Using Nondestructive Evaluation Coupled with Electron Microscopy

**B. Wisner and Antonios Kontsos**

**Abstract** Several fatigue failure modes originate at the microstructural level by the activation, interactions and development of what are referred to as "damage precursors" long before the formation of dominant cracks that grow as a function of loading and crystallographic parameters. In this context, this work presents new developments of an in-house developed experimental mechanics approach to evaluate aspects of microstructure evolution and identify validated damage precursors that are active during fatigue loading by combining Nondestructive Evaluation (NDE) methods with ex situ and in situ Scanning Electron Microscopy (SEM). The used NDE methods include real time Acoustic Emission (AE) monitoring from inside the SEM chamber and Digital Image Correlation (DIC) for strain evolution directly at the grain scale. The coupling between quasi in situ microscopy with actual in situ nondestructive evaluation falls into the ICME framework and the idea of quantitative data-driven and multiscale characterization of material behavior. To demonstrate this approach, Aluminum 2024-T3 specimens were tested using a SEM mechanical testing stage under low cycle fatigue to identify and validate the presence of damage precursors, while correlating their presence with specific parameters extracted by the available NDE data. The reported results show how load information could be correlated with both AE activity, DIC strain maps, and microscopic observations of precipitate fracture and microcracks.

**Keywords** Damage precursors • Inclusion fracture • Particle fracture • Damage detection • In situ • NDE

## 1.1 Introduction

Aluminum alloys are widely used both because of their high strength to weight ratio as well as the existence of large-quantity and controllable quality manufacturing procedures, developed over the past several decades [1–6]. Given their mass adoption, there is a wide variety of examples in engineering applications in which fatigue and fracture of such alloys becomes a dominant economical parameter, as this relates to life cycle related decisions for existing structures and their continuing use in demanding operation environments encountered for instance in marine and aerospace applications, as well as a crucial design parameter for the next generation of lightweight structures starting from physical metallurgy all the way until final product finish. Given this context, the objective of this article is to investigate the role of state-of-the-art and multiphysics nondestructive evaluation (NDE) based on acoustics, optics and microscopy coupled with direct measurements of microstructural effects in material behavior to detect, identify and analyze the contribution of specific microstructural features on the room temperature fatigue behavior of commercial-grade aluminum alloys used in aerospace.

For this reason this research focuses on the aircraft grade Al alloy, 2024-T3. This alloy is a precipitate-hardened material containing Cu, Fe, Si, Mg, and Mn as the primary alloying elements [1]. The specific alloying and temper process results in Cu and Fe rich hard inclusions/particles with sizes $>1$ μm that have been reported in literature since 1969 by Grosskreuts and Shaw to be dominant fatigue crack nucleation sites [7]. In addition, Gao et al. identified similar hard inclusions in Al 7075-T6 and 2024-T3 [8]. Related work by Agarwal et al. attempted to connect bulk strain levels with the number of fractured particles based on surface observations [9]. However, to accomplish this task, interrupted tests were performed to allow for surface observations of damaged particles inside a scanning electron microscope (SEM). Therefore for both material synthesis and damage prevention purposes, it becomes crucial to identify and develop a method capable of informing when surface and subsurface particles fracture and what is their effect on local and global mechanical behavior. In this direction, a number of papers have been published further supporting crack nucleation at hard inclusions [3, 10–12].

B. Wisner • A. Kontsos (✉)
Department of Mechanical Engineering and Mechanics, Drexel University, Philadelphia, PA 19104, USA
e-mail: akontsos@coe.drexel.edu



A.T. Zehnder et al. (eds.), *Fracture, Fatigue, Failure and Damage Evolution, Volume 8*,
Conference Proceedings of the Society for Experimental Mechanics Series, DOI 10.1007/978-3-319-42195-7_1

In this framework the innovation recently proposed by the authors [13] is related to what is termed here as microstructure-sensitive Acoustic Emission (AE) monitoring for identifying damage directly at the scale that this occurs in Aluminum alloys. In fact, the term "acoustic emission" is concurrently used to denote both the natural phenomenon [14–17] that occurs in all materials at multiple scales that is related to the energy release that occurs as a result of mostly irreversible processes such as plasticity and fracture [18], as well as the versatile and passive NDE technique that detects travelling elastic waves by the use of a variety of transduces and signal processing methods [19, 20].

Therefore, AE is suitable for application both at the microstructural and structural levels allowing for connections between fundamental science observations and practical applications. Indeed since the work by Hamstad et al. [21] efforts have been made to distinguish the different sources of AE included slip as well as noise, however, such distinction was difficult if not impossible due to limitations in both observational means (e.g. in situ microscopy) as well as lack of advanced digital signal processing capabilities. Despite these deficiencies, Hamstad and others argued that burst AE signals, similar to those emitted during particle fracture are easier to identify, primarily due to their high amplitude and energy as well as their narrowband frequency content, and therefore link them to the activation of actual damage sources. Moreover, McBride et al. [22] attempted to make a quantitative connection between AE amplitude and particle fracture using Al7075 containing particles and comparing it to pure Al with no particles; AE from particle fracture in this investigation was reported to be burst in nature and resulting from large particle (~10 μm), which the authors supported by providing post mortem evaluation of the specimens. Further work by Wells et al. [23] attempted to separate the dislocation motion (slip) related AE sources from particle fracture using the peaks in root-mean-square (rms) value of the AE signals. The hypothesis is such analysis was that particle fracture only happens in tension while slip occurs in both tension and compression. Therefore these authors argued that signals present in both directions of loading are representative of slip and the signals present only in the tension tests are most likely occurring due to particle fracture. Furthermore, AE has been used to identify damage and material degradation of a variety of materials; in a recent example Ai et al. [24] proposed the use of NDE methods, primarily AE to detect micro-defects and cracks in Al A356 by proposing the use of additional AE features such as counts and energy to determine the level of damage in the material.

The research discussed in this paper proposes to make use of recent advances in technology to perform experiments inside an SEM microscope for direct observation of micro processes of interest (particle fracture) while recording and monitoring AE events in real time. Furthermore, other NDE techniques such as digital image correlation (DIC) are implemented to assist in the identification of specific time, load and strain values of a priori identified particle fractures to directly correlate recorded AE signals with observable fracture events. Specific AE features are extracted and compared to other AE information obtained from other micro processes in both monotonic and fatigue tests. The overall experimental mechanics approach allows the extraction of parameters that can be leveraged by both crystal and continuum level plasticity models that are concurrently developed by the authors.

## 1.2 Experimental Procedure

### *1.2.1 Mechanical and Nondestructive Evaluation Setup*

To combine AE signals, microscale DIC images and microstructure evolution while a specimen is undergoing load the experimental setup shown in Fig. 1.1 was used [13]. Tensile loads were applied by a GATAN MTEST2000 load stage capable of supplying both monotonic tension and compression loads as well as performing cyclic tension–tension and compression–compression fatigue tests. For this work, the stage was placed inside a FEI XL30 ESEM under high vacuum and specimens were loaded at a rate of 0.1 mm/min to allow for in situ monitoring and video recording of microstructure evolution without significant blurring. Additionally, high resolution secondary and backscatter electron images were obtained prior to, and after loading to characterize the materials initial and final state and quantify any induced microstructural changes, as well as to perform 2D DIC deformation measurements.

Acoustic Emission signals were monitored and recorded throughout loading using a four-channel data acquisition system consisting of two PCI-2 boards (manufactured by MISTRAS, Inc.). Two piezoelectric sensors (MISTRAS, Pico Sensor) were attached to the back of the specimen and the signals were passed through the SEM chamber by means of specially made BNC connectors mounted on the MTEST stage door. The signals were obtained at a sampling rate of ten million samples per second to avoid the possibility of signal aliasing. Furthermore, the signals were amplified by using 2/4/6 preamplifiers (MISTRAS, Inc.) and were band-pass filtered in the frequency range of 0.2–1 MHz (which is the manufacturer's defined useful range of the PICO sensors used). The peak definition, hit definition, and hit lock out times were set to 100, 400, and

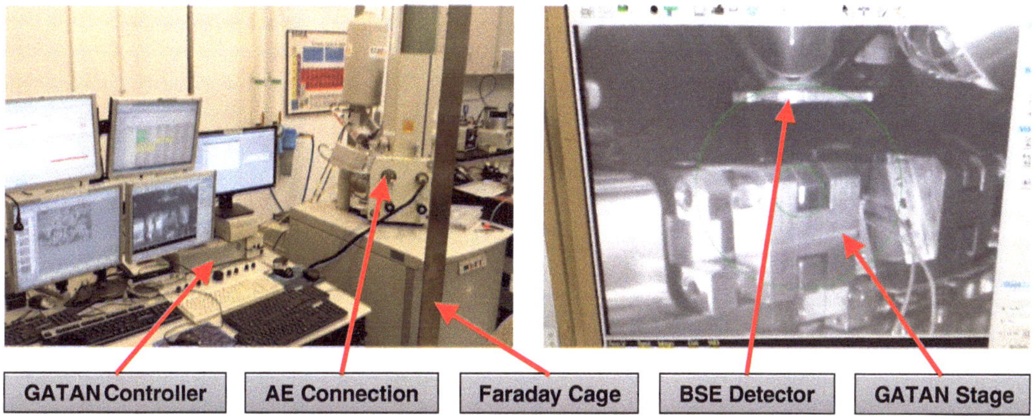


**Fig. 1.1** Experimental set up used consisting of a GATAN MTEST 2000N load cell stage, custom door with BNC connectors for up to four AE channels, and Faraday Cage for EMI reduction in AE signals

500 μs respectively based on the signals obtained in previous tests by the authors [13]. Additionally, the entire SEM and AE acquisition system was contained within a Faraday cage designed to reduce the effects of EMI present in the area on the AE signals obtained. In fact, the presence of this cage was capable to reduce the noise floor from 35 to 20 dB allowing for the acquisition of lower amplitude signals (such as those expected from particle fracture) to be captured and analyzed.

DIC is performed using images obtained from the SEM during loading. The speckle pattern is achieved using Cu particles with average diameter of 300 nm applied to the surface of the specimen through a filter [25]. This method allows for the microstructural changes to be observed though the pattern. Images are analyzed using VIC-2D from correlated solutions to calculate strain localizations. A subset size and step size of 400 μm and 15 μm respectively for the low magnification DIC. High magnification (1000×) resulted in a subset size of 10 μm and a step size of 0.7 μm.

### *1.2.2 Specimen Preparation*

A commercially available 2.5 mm-think rolled Al 2024-T3 sheet was obtained; 28 mm long double notch specimens were cut by EDM from the sheet as shown in Fig. 1.2a. The sharp notch region was designed to localize plasticity for observations in the SEM. Prior loading, the specimens were ground flat and polished to 0.05 alumina suspension for Electron Backscatter Diffraction (EBSD). A representative map is presented in Fig. 1.2b showing grains of ~100 μm mixed with smaller grains on the order of 30 μm. Given the EBSD results in Fig. 1.2b it was concluded that there is no observable texture in the as-received material.

Al 2024 primary alloying elements include Cu, Fe, Mg, Mn, and Si. Consequently, metallic inclusions with sizes greater than 1 μm are abundant. These inclusions/particles have been shown to lead to the formation of fatigue cracks caused both by elastic mismatches between the surrounding matrix and the particles, as well as by geometric discontinuities due to the particles shape. Such effects cause either debonding of the particle from the surrounding matrix [3, 26] and/or particle fracture [3, 10–12]. To determine the composition of these particles and identify possible candidates for fracture, Energy-Dispersive X-ray Spectroscopy (EDS) maps were produced (Fig. 1.2c) since hard inclusions typically have high concentrations of Fe and Cu.

## 1.3 Results and Discussion

Two sets of experiments were conducted to investigate the detection of fatigue damage precursors. Specifically, specimens were initially loaded monotonically up to a point that the first signs of damage prior to crack nucleation appeared. Then, monotonic tests were conducted until several surface particles were seen to fracture to determine the associated AE signals. Finally, this information was investigated by performing cyclic tension-tension testing.

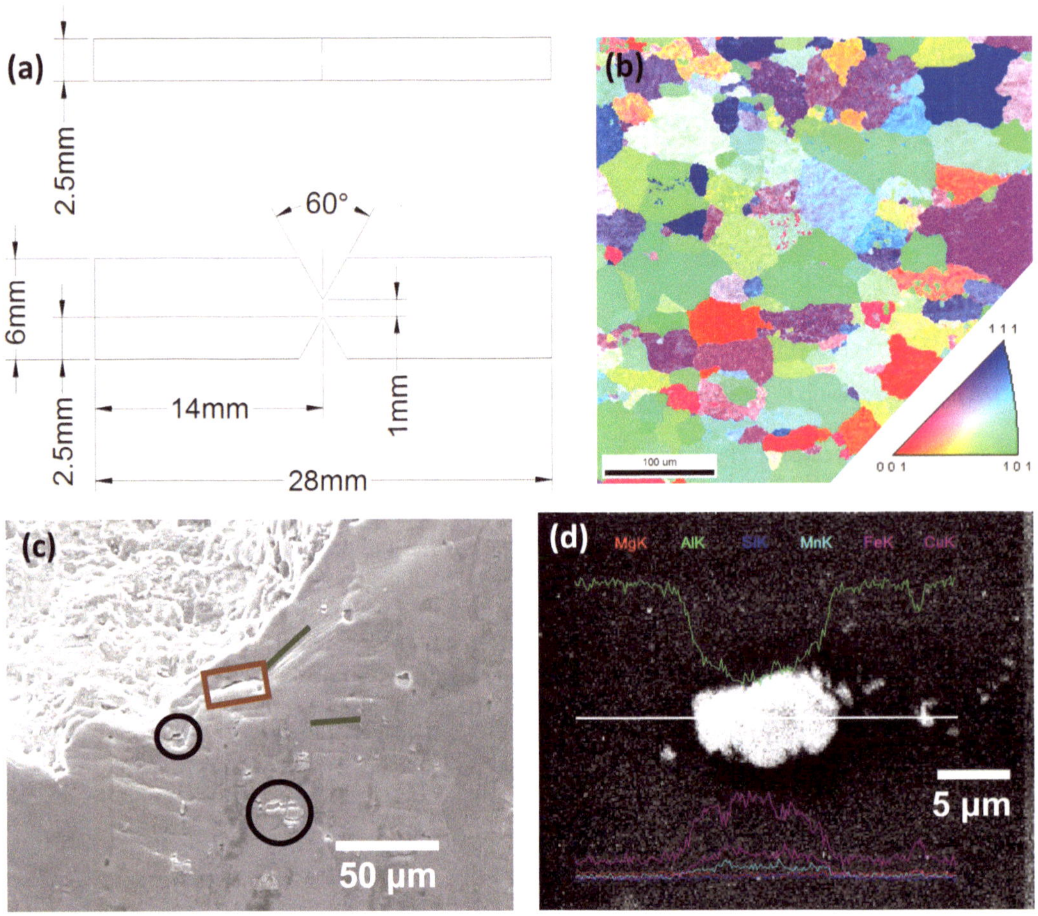


**Fig. 1.2** (**a**) Specimen Geometry with notches to localize the plastic zone. (**b**) EBSD map of the region showing grains on ranging from 20 to over 100 μm. (**c**) Notch region after loading indicating slip (*lines*), void nucleation (*boxes*), and particle fracture (*circles*) as initial damage prior to failure. (**d**) Single particle of interest depicting high Fe and Cu content (hard inclusion)

### *1.3.1 Monotonic In Situ SEM Testing*

Based on the related literature, the relatively larger and hard inclusions tend to be the particles that fracture first [7, 9, 22]. Therefore, an area near the notch root with a number of inclusions with sizes greater than 1 μm was selected for monitoring in the SEM. Secondary Electron imaging was used with frame averaging of four images to reduce the effects of noise without causing blurriness due to the motion. Low frequency continuous signals were recorded continuously as shown in Fig. 1.3a. These AE signals obtained during the stages of first application and subsequent initial stages of loading generally served the purpose of identifying the characteristics of noise (due to both hardware and setup issues) and perhaps some early plasticity which were definitely though distinct with respect to later observed and possibly particle fracture related AE content. Specifically, burst-type signals started to appear later in the load cycle, usually slightly before or at loads near yielding loads (Fig. 1.3a). Live SEM monitoring of particles at the magnification and resolution required to see fracture events resulted in a FOV of approximately 300 $\mu m^2$. Consequently, only a select number of particles could be monitored in real time, while any other fractured ones could be inferred from comparison of images before loading and after failure.

The orange line in Fig. 1.3a indicates the point after which particle fracture was observed, given the procedure described herein. In addition, Figs. 1.3b, c show the monitored region before and after this loading point. More than one particle fracture events was observed in these microscopy images, as well as other mechanisms such as void formation and growth. Figure 1.3d highlights some of these mechanisms.

Additional information can be extracted for the AE data shown in Fig. 1.3a that can be potentially used to identify occurring damage. Specifically, in Fig. 1.4a there are two locations of absolute energy jump, the first being at the onset of "high frequency" (in this case higher than 300 kHz) in Fig. 1.4b, while the second occurs later at the same time as observed

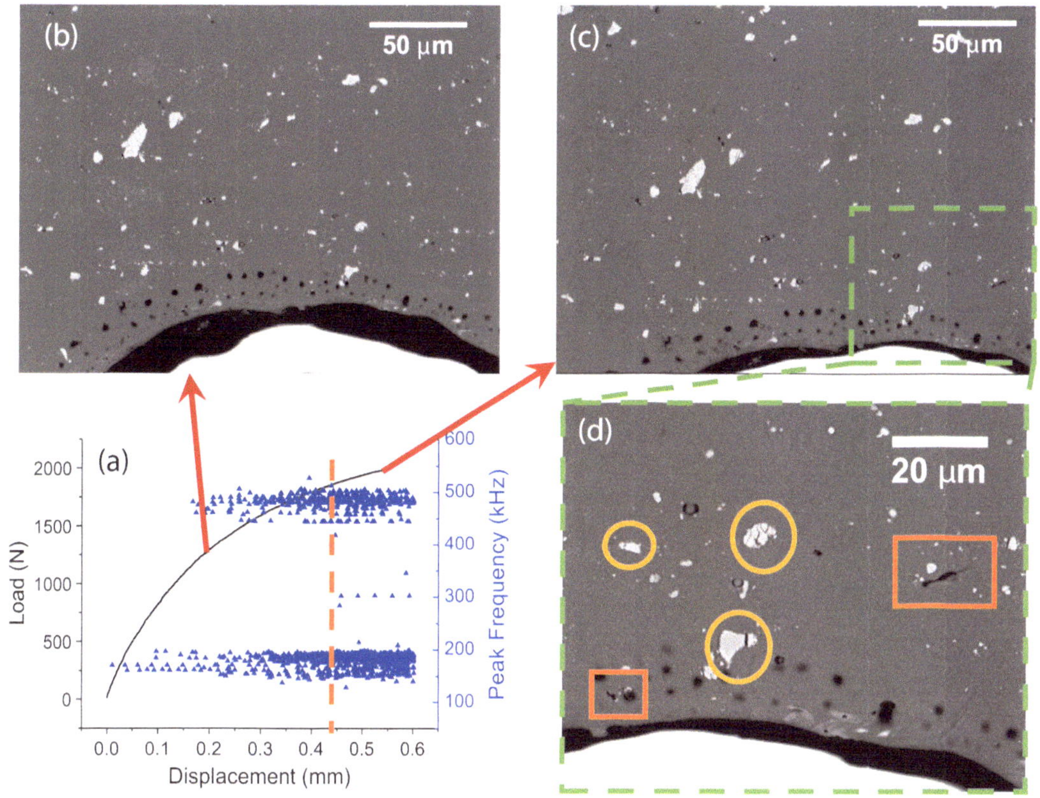


**Fig. 1.3** (**a**) Load profile with AE Peak frequency overlaid. The vertical line indicates the location of observed particle fracture. (**b**) SE micrograph of the monitored region prior to particle fracture. (**c**) SE micrograph of the monitored region after loading stopped because of observed particle fracture. (**d**) Higher magnification image showing particles fractured (*circles*) and voids growing (*boxes*)

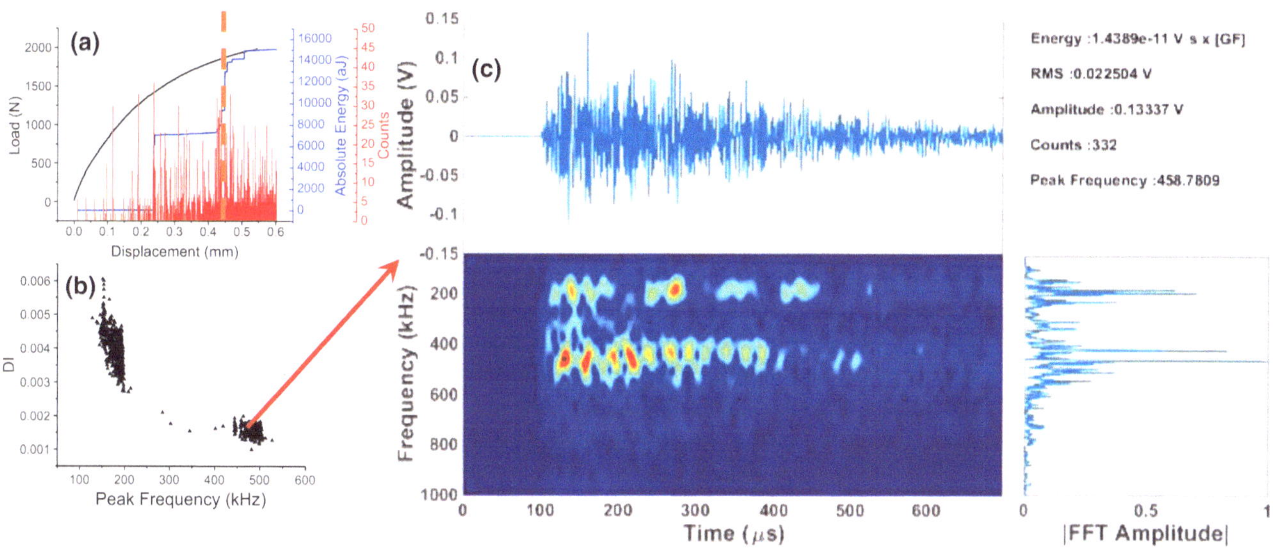


**Fig. 1.4** (**a**) Load profile with cumulative AE Absolute Energy overlaid showing a jump in energy at the time of observed particle fracture. Again, the vertical line indicates the location of observed particle fracture. (**b**) AE Damage Indicator plotted against Peak Frequency showing a cluster of signals with High Frequency and low DI. (**c**) The AE signal that occurred at the time of observed particle fracture plotted with SFFT and FFT to show frequency evolution

particle fracture suggesting damage signals between 450 and 550 kHz signals. These jumps occurred because of the monitoring of sudden increase in AE activity, captured by the spikes in the counts bar plot in Fig. 1.4a. Taking advantage of the real time recorded video, the time and displacement of the observed particle fracture can be estimated. Knowing that particle fracture is itself a sudden event possibly resulting in burst-type acoustic emissions, efforts were made to identify the

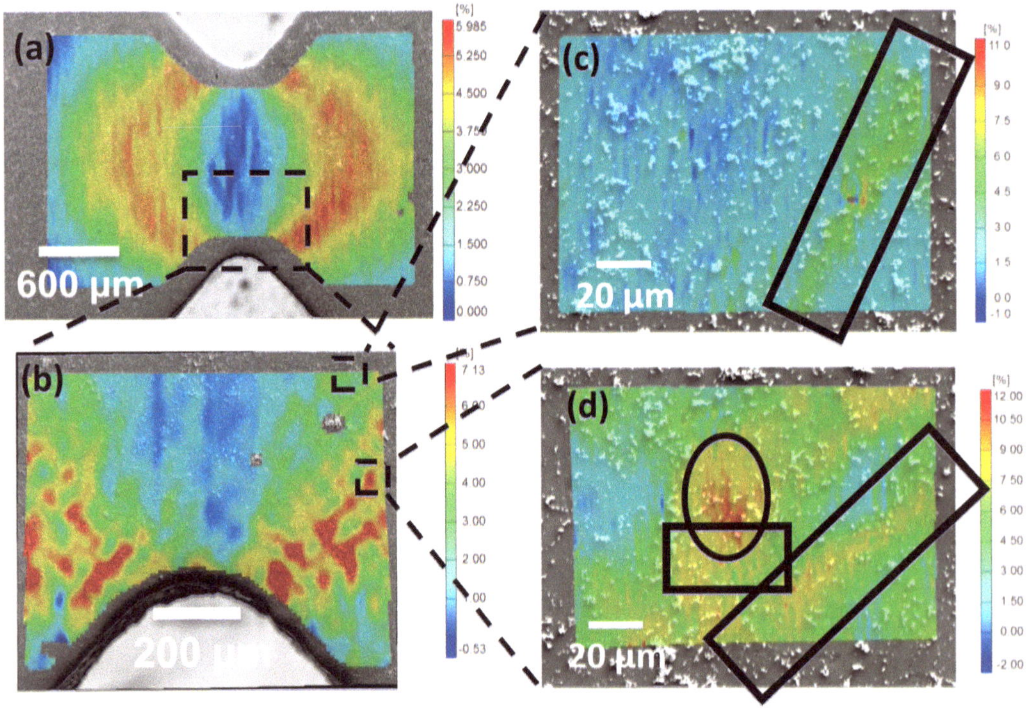


**Fig. 1.5** (**a**) Full Field X direction Strain profile (Specimen was loaded in the X direction) showing geometrically induced strains concentrations as semi-circles. (**b**) Higher magnification DIC was performed revealing the "continuous" strain field (**a**) contains further localizations. Further magnification reveals Strain localizations around *Slip lines* (**c**) as well as around pre-existing voids (**d**). Locations of slip are represented by *boxes* and the void location is identified by the *circle*

specific AE signals resulting from that source (Fig. 1.4c). To this aim, Short time Fast Fourier Transfer (SFFT) results in Fig. 1.4d reveal the frequency content of the signal distributed over time and shows multiple bursts of ~500 kHz throughout the duration of the AE signal in Fig. 1.4c. This observation may suggest that there may be multiple AE signals from multiple sources activated at the same time window of a few 100 ms which then were combined into one AE signal due to the particular hardware and software settings. On-going work by the authors (not reported here) utilizing continuous and not hit-driven waveform streaming has the potential to determine if AE signals such as the one reported in Fig. 1.4c represent one or possible multiple fractures. Additionally, the AE activity can be used to identify key features for determination of a damage indicator, such as the one depicted in Fig. 1.4b based on previous work of the authors [13, 19] which could then be used for possible live detection of particle fracture in both monotonic and fatigue loading conditions.

The concurrent use of other NDE methods such as DIC could add further information and assist in the identification of conditions, e.g. strain localizations that lead to particle fracture and increase, therefore, the reliability of damage source identification using the AE method. In this context, Fig. 1.5 shows related DIC measurements made for this type of specimen. It can be seen using multiscale deformation measurements that "macroscopic" plasticity (Fig. 1.5a) near the notched regions results from "microplasticity" effects (Fig. 1.5b) manifested by the pattern strain localizations radiated in distinct directions from the notch that either induce particle fracture or result from it. The strain measurements in Fig. 1.5b resemble dislocation-related cell patterns observed in aluminum alloys. In addition, localizations around voids (Figs. 1.5c, d) can be seen to combine themselves and result in the localizations seen in Fig. 1.5a.

### *1.3.2 Fatigue In Situ SEM Testing*

The same methodology was used for cyclic testing for a maximum of 45 cycles at 85 % of the measured ultimate strength of the material with $R = 0.1$. In the tests performed, no change was observed on the surface of the specific FOV used as shown in Fig. 1.6, however similar AE activity to that obtained from particle fracture in monotonic loading was observed from the first cycle possibly suggesting that particles are fracturing somewhere in the material. Further fatigue tests are required with either higher loads or more cycles to identify the source of this AE activity.

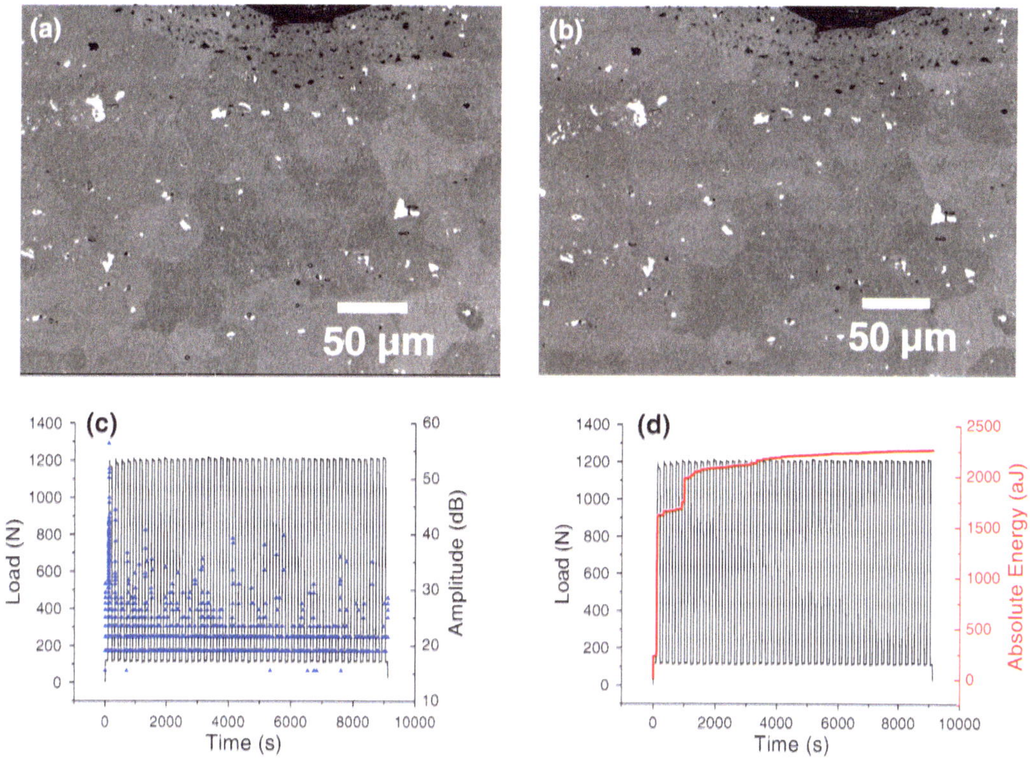


**Fig. 1.6** (**a**) SEM image of the region prior to cyclic loading. (**b**) SEM image of the region after cyclic loading showing no change. (**c**) Load vs time curve with AE amplitude overlaid showing a large number of signals during the first cycle with the number of hits decreasing after. (**d**) Absolute energy curve overlaid over the load curve showing two spikes of energy at cycles one and five suggesting the accumulation of damage in these cycles

## 1.4 Conclusion

Al 2024-T3 fails in fatigue from cracks nucleated from particles fracturing or debonding. NDE methods coupled with in situ SEM observations allowed for the identification of AE sources at the length scale they occur. Specifically, AE data was coupled with information obtained from DIC and changes in the microstructure to identify damage precursors in Al 2024. Particle fracture was connected to representative waveforms, and this information could then be used to quantify a material's state based on the number of signals obtained. The information obtained by AE and DIC could be also fed into models to improve estimates of remaining useful life based on information of the materials physical state.

**Acknowledgements** The authors would like to acknowledge the financial support received by the Office of Naval Research under the Young Investigator Program, Award #N00014-14-1-0571. The authors further acknowledge the technical support received under the National Aeronautics and Space Administration Space Act Agreement, No. SAA1-19439 with Langley Research Center. Finally, the authors acknowledge the technical support received from Sandia National Labs (SAND2016-1174O)

## References

1. Davis, J.R., Davis, J.R.: Aluminum and Aluminum Alloys. ASM International, Materials Park, OH (1993)
2. Esawi, A.M., Aboulkhair, N.T.: Bi-modally structured pure aluminum for enhanced strength and ductility. Mater. Des. **83**, 493–498 (2015)
3. Hahn, G., Rosenfield, A.: Metallurgical factors affecting fracture toughness of aluminum alloys. Metall. Trans. A **6**(4), 653–668 (1975)
4. Hirsch, J., Al-Samman, T.: Superior light metals by texture engineering: optimized aluminum and magnesium alloys for automotive applications. Acta Mater. **61**(3), 818–843 (2013)
5. Saito, Y., et al.: Ultra-fine grained bulk aluminum produced by accumulative roll-bonding (ARB) process. Scr. Mater. **39**(9), 1221–1227 (1998)
6. Vasudevan, A.K., Doherty, R.D.: Aluminum Alloys—Contemporary Research and Applications: Contemporary Research and Applications. Elsevier, Amsterdam (2012)

7. Grosskreuts, J.: Critical mechanisms in the development of fatigue cracks in 2024-T4 aluminum. In: ICF2, Brighton, 1969
8. Gao, M., Feng, C., Wei, R.P.: An analytical electron microscopy study of constituent particles in commercial 7075-T6 and 2024-T3 alloys. Metall. Mater. Trans. A **29**(4), 1145–1151 (1998)
9. Agarwal, H., et al.: Quantitative characterization of three-dimensional damage evolution in a wrought Al-alloy under tension and compression. Metall. Mater. Trans. A **33**(8), 2599–2606 (2002)
10. Gupta, V.K., Agnew, S.R.: Fatigue crack surface crystallography near crack initiating particle clusters in precipitation hardened legacy and modern Al–Zn–Mg–Cu alloys. Int. J. Fatigue **33**(9), 1159–1174 (2011)
11. Payne, J., et al.: Observations of fatigue crack initiation in 7075-T651. Int. J. Fatigue **32**(2), 247–255 (2010)
12. Xue, Y., et al.: Micromechanisms of multistage fatigue crack growth in a high-strength aluminum alloy. Acta Mater. **55**(6), 1975–1984 (2007)
13. Wisner, B., et al.: In situ microscopic investigation to validate acoustic emission monitoring. Exp. Mech. **55**(9), 1705–1715 (2015)
14. Baram, J., Rosen, M.: Acoustic emission generated during the tensile testing of aluminium alloys. Mater. Sci. Eng. **40**(1), 21–29 (1979)
15. Fisher, R., Lally, J.: Microplasticity detected by an acoustic technique. Can. J. Phys. **45**(2), 1147–1159 (1967)
16. Ono, K.: Acoustic emission arising from plastic deformation and fracture. J. Acoust. Soc. Am. **64**, 174–175 (1978)
17. Richeton, T., Weiss, J., Louchet, F.: Dislocation avalanches: role of temperature, grain size and strain hardening. Acta Mater. **53**(16), 4463–4471 (2005)
18. Cuadra, J., et al.: Energy dissipation via acoustic emission in ductile crack initiation. Int. J. Fract. **199**(1), 89–104 (2016)
19. Vanniamparambil, P., Guclu, U., Kontsos, A.: Identification of crack initiation in aluminum alloys using acoustic emission. Exp. Mech. **55**(5), 837–850 (2015)
20. Vanniamparambil, P.A., et al.: Cross-validated detection of crack initiation in aerospace materials. In: SPIE Smart Structures and Materials + Nondestructive Evaluation and Health Monitoring, International Society for Optics and Photonics, San Diego, CA, March 2014
21. Mukherjee, A., Hamstad, M., Bianchetti, R.: Origin of burst-type acoustic emission in unflawed 7075-T6 aluminum. J. Test. Eval. **4**(5), 313–318 (1976)
22. McBride, S., MacLachlan, J., Paradis, B.: Acoustic emission and inclusion fracture in 7075 aluminum alloys. J. Nondestruct. Eval. **2**(1), 35–41 (1981)
23. Wells, R., Hamstad, M., Mukherjee, A.: On the origin of the first peak of acoustic emission in 7075 aluminium alloy. J. Mater. Sci. **18**(4), 1015–1020 (1983)
24. Ai, Y., et al.: Investigation of material performance degradation for high-strength aluminum alloy using acoustic emission method. Metals **5**(1), 228–238 (2015)
25. Carroll, J.D., et al.: High resolution digital image correlation measurements of strain accumulation in fatigue crack growth. Int. J. Fatigue **57**, 140–150 (2013)
26. Gall, K., et al.: On the driving force for fatigue crack formation from inclusions and voids in a cast A356 aluminum alloy. Int. J. Fract. **108**(3), 207–233 (2001)

# Chapter 2
# Experimental Fracture Analysis of Tropical Species Using the Grid Method

**Bernard Odounga, Rostand Moutou Pitti, Evelyne Toussaint, and Michel Grediac**

**Abstract** The fracture processes of three tropical species: *Aucoumea klaineana*, *Malicia excelsa* and *Pterocarpus soyauxii*, are investigated with the grid method. These species are widely used in many sub-tropical countries, in timber building construction, as well as in semi-finished products and paper fabrication. However their fracture behaviour must still be investigated, data being scarcely available on this subject. Modified Mixed Mode Crack Growth specimens are used in order to obtain a stable crack growth evolution in opening, shear and mixed mode ratios. The images of the grid are analysed to provide the crack opening displacement and the crack tip location. The stress intensity factors and the critical energy release rates for each species are then obtained by using the compliance method in imposed displacement. The semi-experimental energetic method is also applied in order to show the efficiency of the proposed technique to characterize the fracture properties of the tropical species under study.

**Keywords** Mixed mode crack growth • Tropical wood species • Compliance method • Grid method

## 2.1 Introduction

Today, wood is one of the solutions that sustainably limits global warming of our planet. In Central Africa and particularly in equatorial regions, the forest plays a key role in this regulation. In the case of Gabon, the recent decision of stopping the exportation of tropical species has opened the way to their local extension and to their more widespread use in local individual and industrial buildings. However, the mechanical behavior of local species largely depends on temperature and humidity variations, which can impair their structural efficiency, and thus lead to sudden failure during their service life. Studying failure of tropical wood is crucial to know their behavior and guide local people in their choice of building materials other than concrete and steel. This would also contribute to the sustainable development discussed during the COP21 meeting in Paris last year [1].

This paper is focused on the study of the initiation and the crack propagation at room temperature of the following tropical species: Iroko (*Pterocarpus soyauxii*), Padouk (*Malicia excelsa*) and Okume (*Aucoumea klaineana*). A short review of the literature shows that few studies and data related to failure of this type of wood material are available. Similar studies deal with temperate species as Douglas, Maritime pine and Abies Alba mills, using Compact Tension Shear (CTS) specimens [2–4], Double Cantilever Beam (DCB) specimens [5], Mixed Mode Crack Growth (MMCG) [6] specimens and TDCB specimens [7]. In the present work, tropical wood specimens are investigated with the grid method [8–11].

B. Odounga
Université des Sciences et Techniques de Masuku, Ecole Polytechnique de Masuku, BP 901 Franceville, Gabon

Université Clermont Auvergne, Université Blaise Pascal, Institut Pascal, BP 20206, F-63000 Clermont-Ferrand, France

CNRS, UMR 6602, Institut Pascal, F-63171 Aubiere, France

R.M. Pitti (✉)
Université Clermont Auvergne, Université Blaise Pascal, Institut Pascal, BP 20206, F-63000 Clermont-Ferrand, France

CNRS, UMR 6602, Institut Pascal, F-63171 Aubiere, France

CENAREST, IRT, 14070 Libreville, Gabon
e-mail: rostand.moutou_pitti@univ-bpclermont.fr

E. Toussaint • M. Grediac
Université Clermont Auvergne, Université Blaise Pascal, Institut Pascal, BP 20206, F-63000 Clermont-Ferrand, France

CNRS, UMR 6602, Institut Pascal, F-63171 Aubiere, France



A.T. Zehnder et al. (eds.), *Fracture, Fatigue, Failure and Damage Evolution, Volume 8*,
Conference Proceedings of the Society for Experimental Mechanics Series, DOI 10.1007/978-3-319-42195-7_2

The first part of this paper presents the materials and methods applied for this experimental campaign. The background of the grid technique in order to record the crack tip position during the test is recalled. The origin and the characteristics of the tropical wood species are also described. The experimental devices are detailed. Experimental results are given in term of force-displacement curves and the crack tip location. The second part shows the fracture analysis by computing the critical energy release rate at the failure point using the compliance method.

## 2.2 Materials and Method

### 2.2.1 *Wood Specimen*

The dimensions of the CTS specimens are 105 × 105 × 12.5 mm$^3$ (see Fig. 2.1a). A typical Padouk specimen is shown in Fig. 2.1b. The main difference between the three species is their density: Okume (density = 0.44) is less dense than Iroko (density = 0.64) and Padouk (density = 0.79). It means that during cracking process, Okume exhibits a quasi-brittle behavior. For all the specimens, the initial crack length $a = 25$ mm is at mid-height and oriented along the fiber direction. The initial crack is completed by a notch (length: 3 mm) with a cutter in order to initiate correctly crack propagation. A grid, with a regular pitch of 200 μm, is transferred on one face of the specimen (see Fig. 2.1b).

### 2.2.2 *Devices of the Experiments*

The experimental device is shown in Fig. 2.2. A 200 kN Zwick/Roel testing machine was used for the tests. A camera was fixed on a tripod at a distance of 67.5 cm from the specimens in order to shot grid images during the tests.

A miniature steel Arcan fixture was used to load the specimens. The grips were equipped with a fixed lower frame. A mobile upper part allowed to apply the Arcan system in various mixed mode configurations. The testing machine, driven with imposed displacements, was equipped with force and displacement sensors in order to obtain force-displacement curves. The cross-head speed was 0.005 mm/s, and the acquisition rate of the camera was 1.35 frames/s. To minimize image noise, each image was averaged with 8 frames. The sample was illuminated by three flexible optical arms powered by a cold light source KL 2500 LCD. A SIGMA 105 mm objective was mounted on the camera.

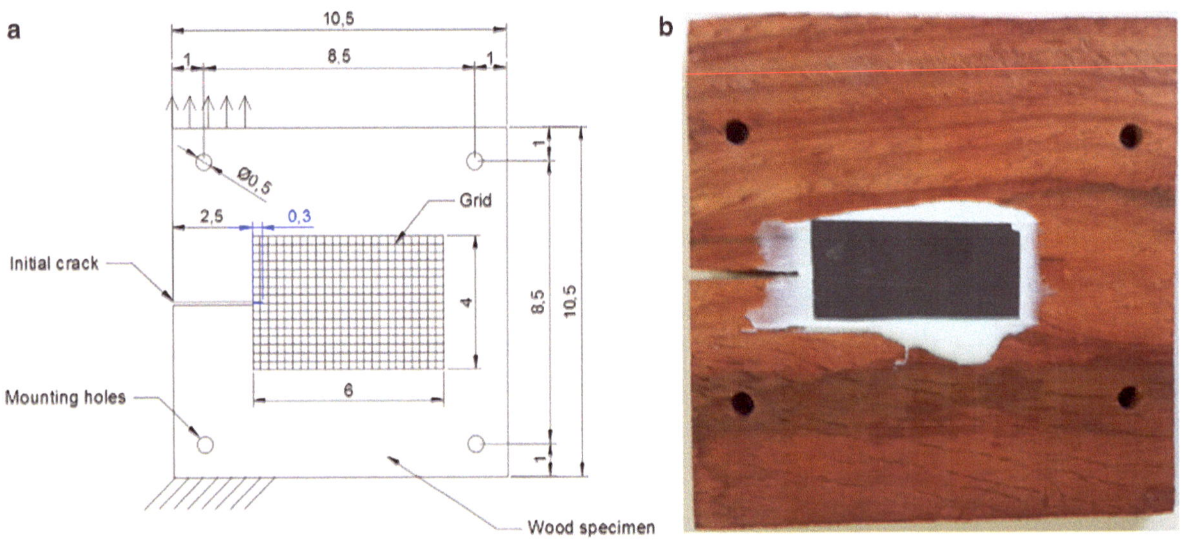


**Fig. 2.1** (**a**) Dimensions of the wooden specimen fitted grid; (**b**) wooden specimen with grid (Padouk)

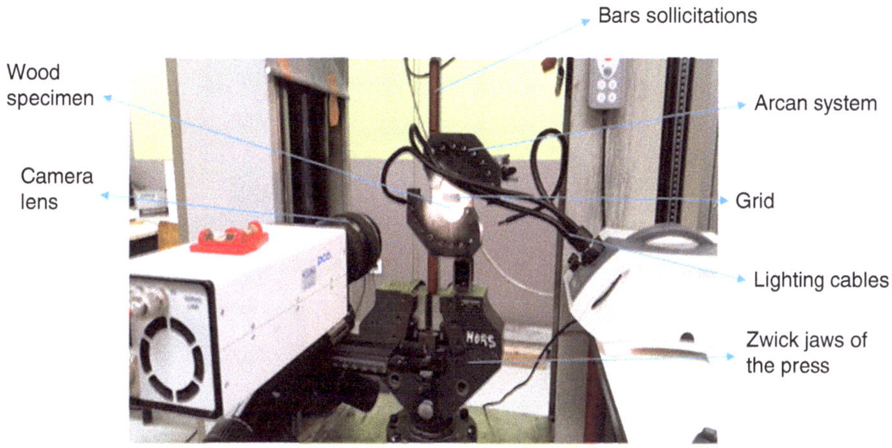


**Fig. 2.2** Experimental device

## 2.3 Results

### 2.3.1 *Force-Displacement Curves*

Figures 2.3, 2.4 and 2.5 show the force-displacement curves for Okume, Iroko and Padouk species, respectively. Note that the fracture toughness is not considered here due to the sudden failure observed for the CTS specimens. Indeed, for this kind of specimen, we do not observe a crack growth process, thus the critical force $F_{CO}$ is equal to the load at failure $F_{RO}$. The crack growth observed here is only due to the uncohesion of wood fibers at the tip. For Okume, the force at failure is equal to $F_{RO} = 147\,\text{N}$, which corresponds to a global vertical displacement $d_O = 0.35\,\text{mm}$. The sudden decrease of load represents the dissipation energy during crack propagation. Due to its low density, failure occurs suddenly.

In the case of Iroko (see Fig. 2.4), the failure force is equal to $\text{F}_{\text{RI}} = 375$ N, which corresponds to a global vertical displacement of Iroko $\text{d}_\text{I} = 0.39\,\text{mm}$. Padouk, Fig. 2.5, has the most important fracture toughness: $\text{F}_{\text{RP}} = 569\,\text{N}$, which corresponds to the maximal displacement of Padouk $\text{d}_\text{P} = 0.69\,\text{mm}$. Before failure, we can see some energy dissipation at about $\text{d}_\text{I} = 0.39$ and 0.65 mm. According to the high density of this specimen, the stable crack growth process is observed after the fracture point, between the displacements of Padouk $\text{d}_\text{P} = 0.75$ and 0.85 mm.

### 2.3.2 *Crack Length-Images Curves*

Figures 2.6, 2.7 and 2.8 show the evolution of the crack vs. image number (and thus time) for Okume, Iroko and Padouk, respectively. Okume presents a sudden crack growth due to instantaneous failure corresponding to the maximum range of $a = 11$ mm (see Fig. 2.6). However, the observation of the Iroko curve (see Fig. 2.7), shows that the maximum crack length is $a = 16$ mm after 350 images. Note that this phenomenon is not only due to the bridging fibers during the increase of the load, but also to the peculiar properties of tropical species. Padouk (see Fig. 2.8), exhibits the maximum crack growth, with about $a = 32$ mm (image #285). In this case, the sudden bridging of the fibers is very marked at about image #240.

## 2.4 Fracture Analysis

In this part, the critical value of the energy release rate of the three species was calculated using the compliance method, Eq. (2.1). These results are calculated with the force at failure, and with the corresponding value of the crack length $a$

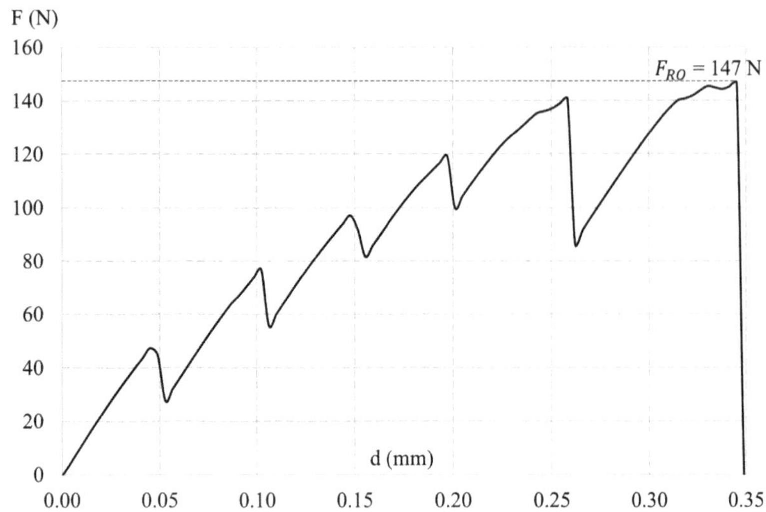


**Fig. 2.3** Force-displacement curve for Okume ($F_{RO}$ = failure force of Okume)

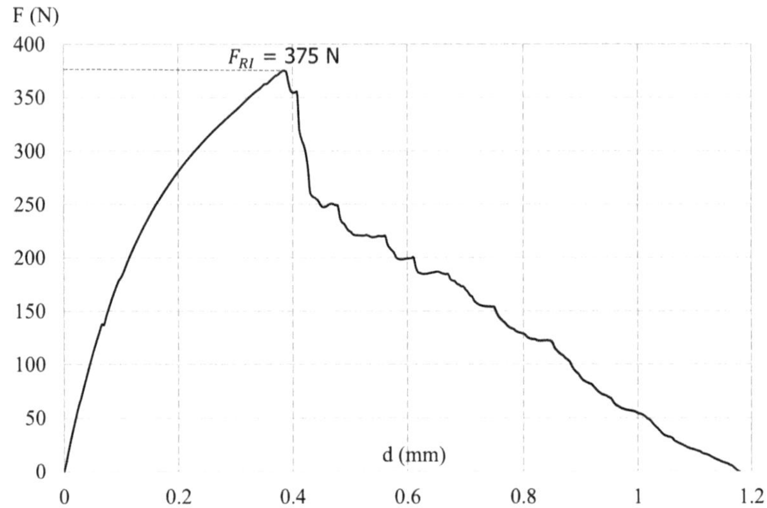


**Fig. 2.4** Force-displacement curve for Iroko ($F_{RI}$ = failure force of Iroko)

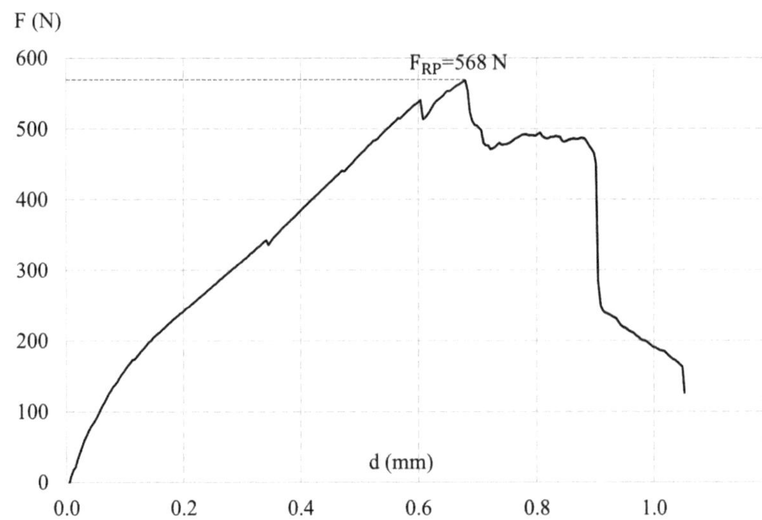


**Fig. 2.5** Force-displacement curve for Padouk ($F_{RP}$ = failure force of Padouk)

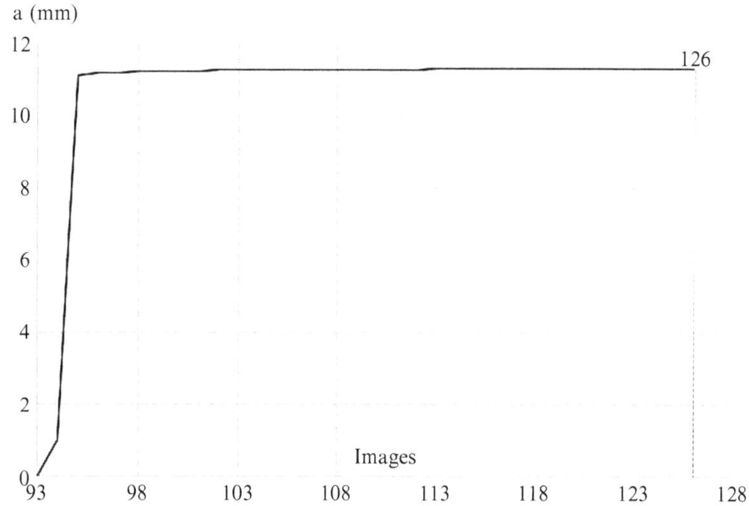


**Fig. 2.6** Crack length—image number for Okume

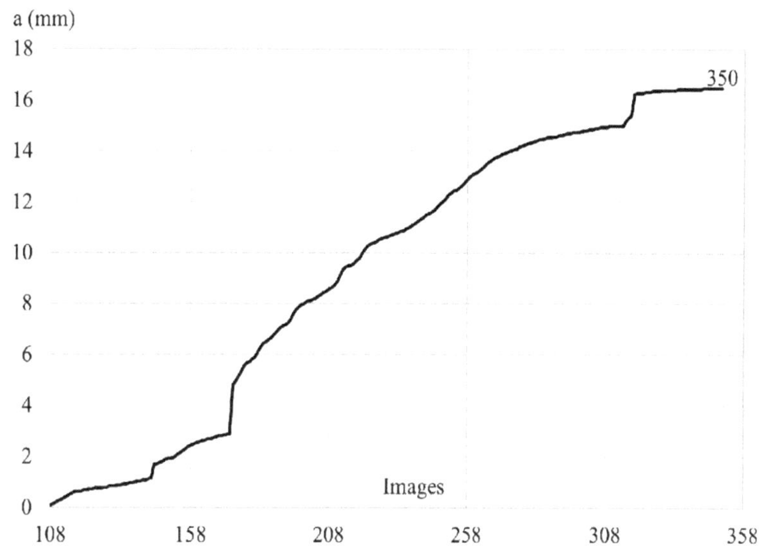


**Fig. 2.7** Crack length—image number for Iroko

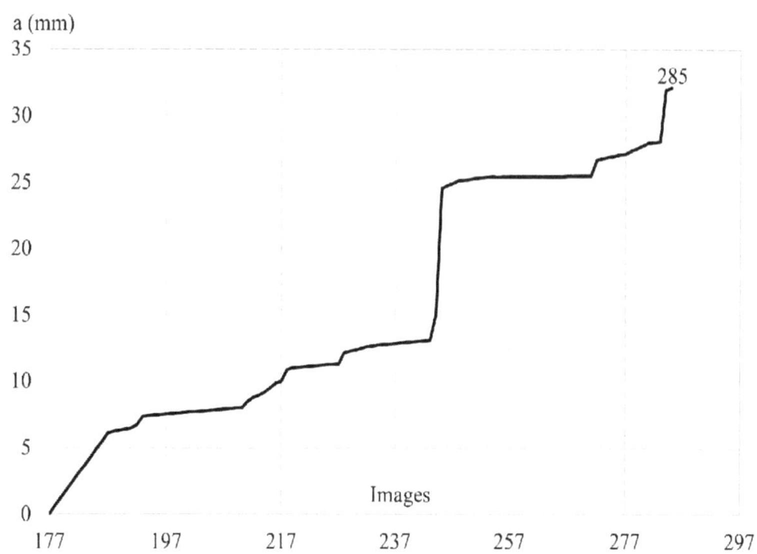


**Fig. 2.8** Crack length—image number for Padouk

**Table 2.1** Critical energy release rate $G_C$

| Wood species | Density | Crack length $a$ (mm) | Compliance $C$ (N/m) | Rupture force $F$(N) | Critical energy release rate $G_C$ (J/m$^2$) |
|---|---|---|---|---|---|
| *Aucoumea klaineana* (Okume) | 0.44 | 28 | 2.38E−06 | 147 | 82 |
| *Milicia excelsa* (Iroko) | 0.64 | 28 | 1.03E−06 | 375 | 234 |
| *Pterocarpus soyauxii* (Padouck) | 0.79 | 28 | 1.22E−06 | 568 | 632 |

$$G_C = \frac{F_C^2}{2b} \cdot \left( \frac{\partial C}{\partial a} \right)_d \tag{2.1}$$

where $F_c$ is the critical load inducing a crack propagation length, $b$ is the thickness of the specimen, and $C$ denotes the compliance at imposed displacement. The results obtained are summarized in Table 2.1.

The values of density given by CIRAD [12] show that Padouk is denser than Iroko and Okume (see Table 2.1). Indeed, Okume is considered as a very light wood, Iroko as medium light wood and Padouk as medium-heavy / heavy wood. The present results show that the higher the density, the higher the value of the energy release rate. Thus the energy release rate value of Padouk is 2.5 times greater than that of Iroko, and that of Iroko is also more than 2.5 times that of Okume (see Table 2.1).

## 2.5 Conclusion

Tests were performed on three wood tropical species. The grid method was applied in order to follow the crack tip evolution during the test in opening mode. The critical energy release rate was obtained with the compliance method. It has been observed that the fracture toughness of Padouk is higher than that Iroko and Okume. In further work, the tests will be performed in mixed-mode configuration and for various orientations of the fibers.

## References

1. Hourcades, J.C., Shukla, P.R.: Cancun's paradigm shift and COP 21: to go beyond rhetoric. Int. Environ. Agreements. **15**(4), 343–351 (2015). ISSN: 1567-9764 (print) 1573-1553 (online)
2. Valentin, G., Caumes, P.: Crack propagation in mixed mode in wood: a new specimen. Wood Sci. Technol. **23**, 43–53 (1989)
3. Angellier, N., Moutou Pitti, R., Dubois, F.: Crack Analysis of Wood Under Climate Variations, vol. 8, pp. 235–242. Springer, Berlin (2016). doi:10.1007/978-3-319-21611-9_29
4. Moutou Pitti, R., Dubois, F., Petit, C., Sauvat, N.: Mixed mode fracture separation in viscoelastic orthotropic media: numerical and analytical approach by the Mtv-integral. Int. J. Fract. **145**, 181–193 (2007)
5. Phan Ngoc, A., Morel, S., Chaplain, M., Coureau, J.L.: R-curve on fracture criteria for mixed-mode in crack propagation in quasi-brittle material: application for wood. Procedia Mater. Sci. **3**, 973–978 (2014). Original Research Article
6. Moutou Pitti, R., Dubois, F., Pop, O.: On a specimen providing stable mixed mode crack growth in wooden material. Comptes Rendus Mécanique **336**(9), 744–749 (2008). Original Research Article
7. Coureau, J.L., Morel, S., Dourado, N.: Cohesive zone model and quasibrittle failure of wood: a new light on the adapted specimen geometries for fracture tests. Eng. Fract. Mech. **109**, 328–340 (2013). Original Research Article
8. Badulescu, C., Grédiac, M., Mathias, J.-D.: Investigation of the grid method for accurate in-plane strain measurement. Meas. Sci. Technol. **20**(9), 1–17 (2009)
9. Dang, D., Toussaint, E., Grédiac, M., Moutou Pitti, R.: Investigation of the Hydric Transfer Phenomenon in Wood at the Ring Scale with the Grid Method, vol. 4, pp. 77–81. Springer, Berlin (2016). doi:10.1007/978-3-319-22449-7_9
10. Piro, J.L., Grédiac, M.: Producing and transferring low-spatial-frequency grids for measuring displacement fields with moiré and grid methods. Soc. Exp. Mech. **28**(4), 23–26 (2004)
11. Sur, F., Grédiac, M.: On noise reduction in strain maps obtained with the grid method by averaging images affected by vibrations. Opt. Lasers Eng. **66**, 210–222 (2015)
12. CIRAD Sources: Centre de coopération International en Recherche Agronomique pour le développement, TROPIX 7- © 1998–2011. http://tropix.cirad.fr/FichiersComplementaires/FR/Afrique/PADOUK.pdf

# Chapter 3
# Investigating the Effective Fracture Toughness of Heterogeneous Materials

**Chun-Jen Hsueh, Guruswami Ravichandran, and Kaushik Bhattacharya**

**Abstract** Heterogeneous materials are ubiquitous in nature, and are increasingly being engineered to obtain desirable mechanical properties. Naturally, the bulk properties of a heterogeneous material can be different from those of its constituents. Thus, one needs to determine its overall or effective properties. For some of these properties, like effective elastic modulus, the characterization is well-known, while for other such as effective fracture toughness, it is a matter of ongoing research. In this paper, we present a method to measure the effective fracture toughness. For the method, we apply a time-dependent displacement condition called the surfing boundary condition. This boundary condition leads the crack to propagate steadily macroscopically but in an unconstrained manner microscopically. We then use the grid method, a non-contact full-field displacement measurement technique, to obtain the displacement gradient. With this field, we compute the macroscopic energy release rate via the area J-integral. Finally, we interpret the effective toughness as the peak of the energy release rate. Using this method, we investigate the influence of heterogeneity on effective fracture toughness. We find that the effective toughness can be enhanced due to the heterogeneity. Consequently, engineered heterogeneity may provide a means to improve fracture toughness in solids.

**Keywords** Fracture toughness • Heterogeneous materials • Surfing boundary conditions • Grid method • Brittle materials

## 3.1 Introduction

Methods for measuring the toughness of homogeneous materials are well-established, but these methods do not extend to heterogeneous materials. While there are established methods for specific heterogeneous materials such as R-curve measurements for ceramics [1] and the double cantilever beam (DCB) method for the laminated materials [2], a general method for materials with complex heterogeneity is wanting. Similarly, a systematic understanding of the effective toughness, crack nucleation, deflection, and pinning in heterogeneous materials, and the dependence of toughness on microstructure are active research topics. Recently, Hossain et al. [3] proposed an approach to define the effective toughness and used it to study the role of microstructure on toughness. Here, we propose an experimental method to measure the effective toughness of heterogeneous materials based on the theoretical work of Hossain et al. [3]. The key idea is to enforce a steady and controlled crack growth at the macroscopic scale without constraining the microscopic scale.

## 3.2 Experimental Configuration

In this study, we developed an experimental configuration that enables macroscopically steady crack growth while allowing at the microscopic scale crack deflection and pinning, nucleation of distal cracks, etc. The experimental setup is shown in Fig. 3.1. A PMMA specimen which is 3.125 mm (1/8″) thick is precisely shaped using a laser cutter. The specimen is connected to rollers which roll on an aluminum rail, and a stepper motor stage pulls the aluminum rail downwards while the specimen is held in place by an aluminum frame. As a result of our choice of the shape of the rails, when the stepper motor pulls the rail down and the specimen slides along, the rail imposes a smoothly translating crack-opening displacement. This is an experimental manifestation of the surfing boundary conditions [3].

C.-J. Hsueh • G. Ravichandran • K. Bhattacharya (✉)
Division of Engineering and Applied Science, California Institute of Technology, 1200 E California Blvd, Pasadena, CA 91125, USA
e-mail: chhsueh@caltech.edu; bhatta@caltech.edu



A.T. Zehnder et al. (eds.), *Fracture, Fatigue, Failure and Damage Evolution, Volume 8*,
Conference Proceedings of the Society for Experimental Mechanics Series, DOI 10.1007/978-3-319-42195-7_3

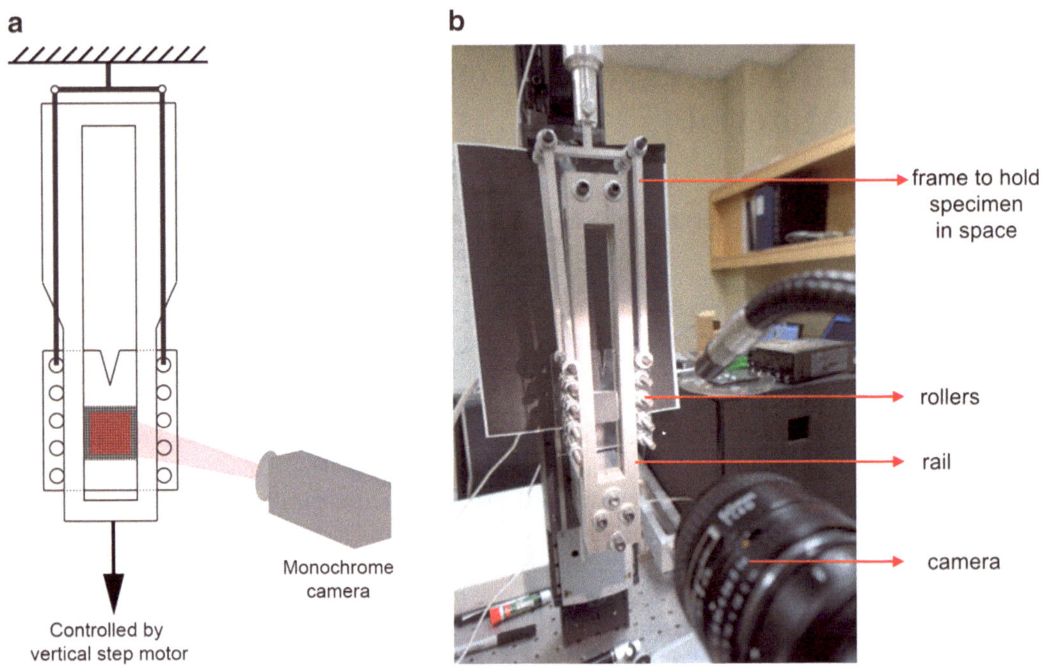


**Fig. 3.1** (**a**) Schematic of the experiment configurations; (**b**) a photograph of the experiment setup

**Fig. 3.2** Schematic of the grid method

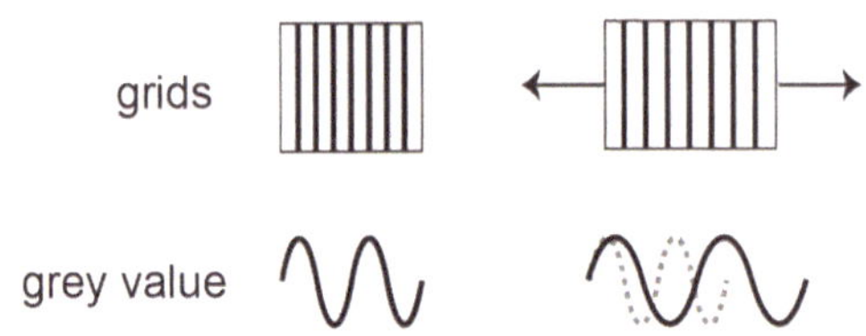


## 3.3 Grid Method

We use a grid pattern to calculate the strain field during the fracture process. The grid method is an optical method that can measure the full-field displacement and displacement gradient [4–7]. A precise grid pattern is carefully applied on the region of interest and the reference and deformed intensities are compared to obtain the desired fields [8]. This method has been widely used in many different research areas [9–13] as it has the advantage that the strain can be directly obtained from the images [12]. In the undeformed state, the grid induces a periodic function on the grey value. When the specimen is deformed, the frequency change in the periodic grey value is directly related to the strain field (Fig. 3.2). The local frequency of the grey value is obtained by the windowed Fourier transform (WFT).

## 3.4 Energy Release Rate

We determine the macroscopic energy release rate by computing the macroscopic J-integral. The J-integral characterizes the driving force for the crack [14]. In order to sample more data points and to reduce the sensitivity to experimental noise, we use the area J-integral instead of the line J-integral [15].

$$J_i = -\int_A \left(u_{k,i}\sigma_{kj} - W\delta_{ij}\right)q_{,j}\, dA \tag{3.1}$$

where $q$ is an arbitrary function which takes value 1 at the outer boundary of the integral area and 0 at the inner boundary. In calculating the $J$-integral, we use the grid method to obtain the displacement gradient field. However, the grid method cannot handle the displacement jump and an artificially large displacement gradient is created in a region near the displacement jump. To eliminate this effect, we truncate the area $J$-integral near the crack face.

## 3.5 Results

### *3.5.1 Homogeneous Case*

To validate our method, we first characteristize the fracture toughness of a homogeneous PMMA specimen. The displacement gradient field measured by the grid method is smoothed by a Gaussian filter with size 3 × 3 and standard deviation 1. Since the grid method creates artificially large displacement gradient near the crack face, we truncate the integration area in this regime with a threshold normal strain equal to 0.01. After calculating the J-integral, we obtain the critical stress intensity factor by Irwin's formula,

$$J = \frac{K_I^2}{E}. \tag{3.2}$$

The value of the deformation gradient and the integration area for the J-integral, which is truncated near the crack face are shown in Fig. 3.3a. Figure 3.3b shows that the stress intensity factor is nearly constant and the value is in the region of published values of PMMA [16] ~ 1 MPa$^{1/2}$.

### *3.5.2 Heterogeneous Case*

We fabricated the heterogeneous specimens by using a laser to engrave grooves to make weak regions. In the first specimen, we made parallel weak lines in the region of interest. The weak lines are shown in Fig. 3.4a with red arrows. Since Irwin's formula (Eq. 3.2) cannot be used on heterogeneous materials, we use the J-integral away from the heterogeneity to characterize the effective driving force of the crack. As Fig. 3.4b shows, the value of the macroscopic energy release rate (J-integral) fluctuates when the crack propagates in the weak lines region.

We then made the engraved grooves deeper, thus increasing the heterogeneity. We repeated the experiment using the same experimental setup (Fig. 3.1). For higher heterogeneity, the crack bifurcates and then gets trapped after the first weak line. The crack path is shown in Fig. 3.5a while Fig. 3.5b shows the strain field for the bifurcated crack and Fig. 3.5c shows the J-integral. We can see that when the crack gets trapped, the value of J-integral increases dramatically. Therefore, the bifurcation and pinning by the weak line can be viewed as a toughening mechanism due to material heterogeneity.

We further increased the depth of the groove and we also increased the width of the weak line. Repeating the experiment with this heterogeneity, we found that the crack bifurcated after the first weak line similar to the previous result. However, in loading the specimen further, the bifurcated crack remained trapped but a new crack nucleated on the weak plane and was subsequently deflected. After some time, the new crack became trapped and another new crack nucleated. As a result, we observed several inclined cracks after the weak line as shown in Fig. 3.6a. The corresponding strain map is shown in Fig. 3.6b. Similar to the previous result, the J-integral increased greatly when the crack was trapped (Fig. 3.5c).

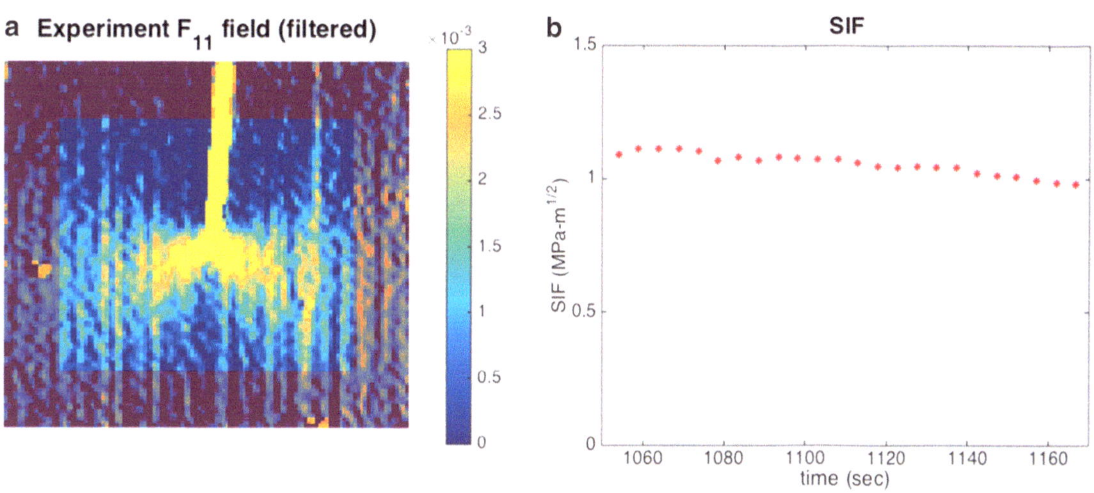


**Fig. 3.3** (**a**) Displacement gradient field and integration area for *J*-integral; (**b**) critical stress intensity factor for homogeneous PMMA as a function of time during steady crack propagation

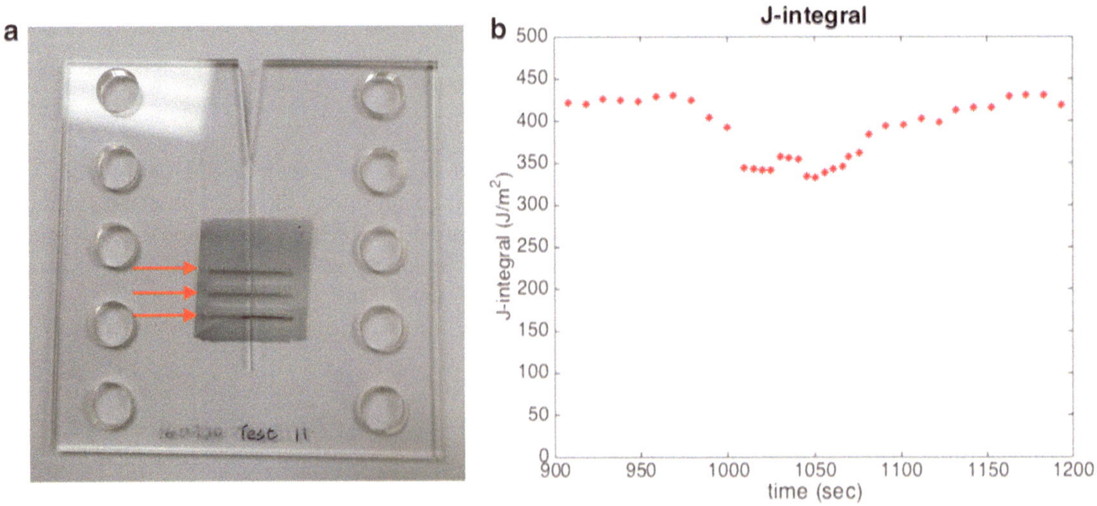


**Fig. 3.4** (**a**) Heterogeneous specimen with periodic surface grooves and (**b**) corresponding energy release rate ($J$) during crack propagation

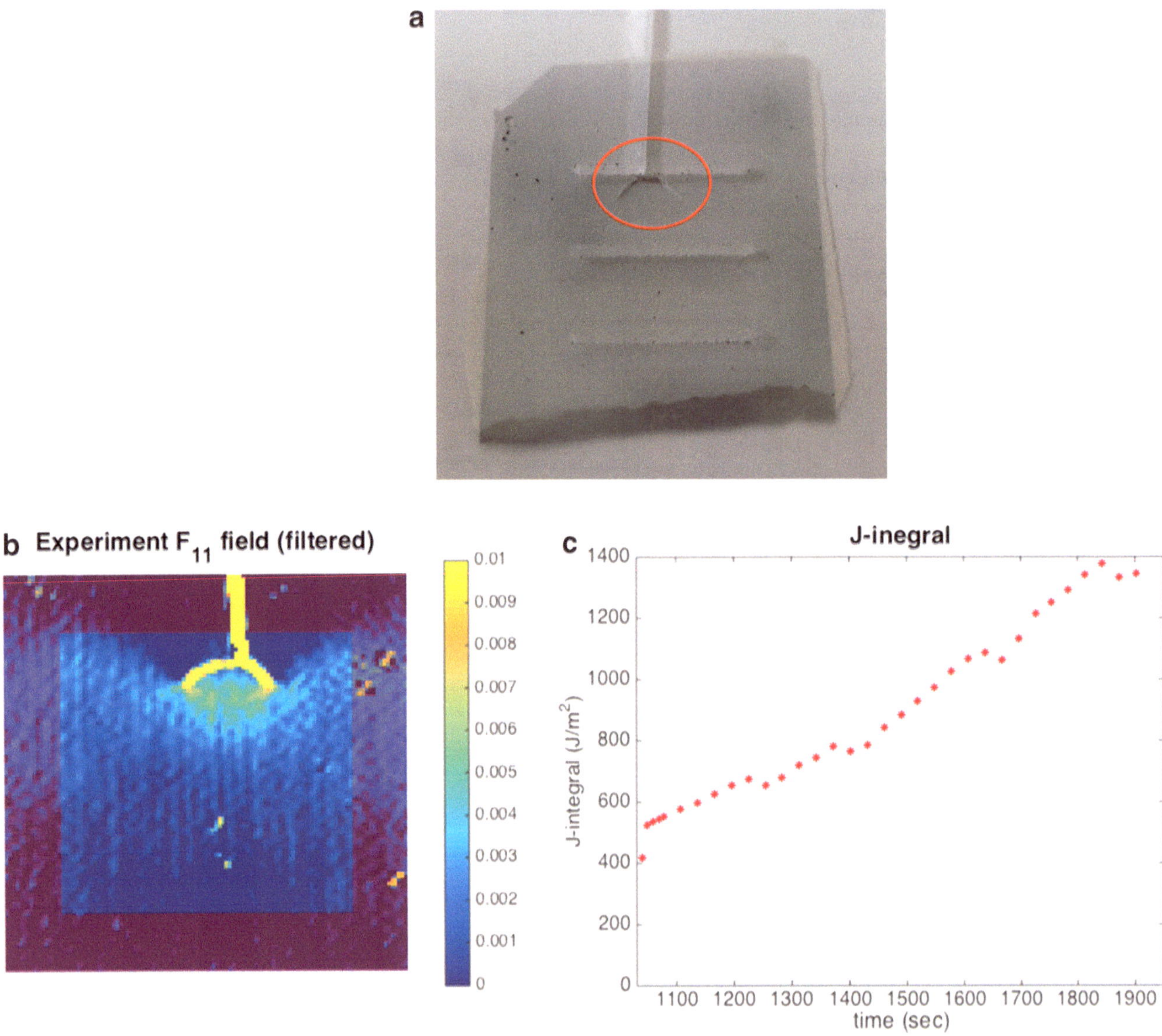


**Fig. 3.5** (**a**) Photograph of bifurcated crack after being trapped at the weak plane, (**b**) deformation gradient map and integration area for $J$-integral, (**c**) $J$-integral as a function of time during crack propagation

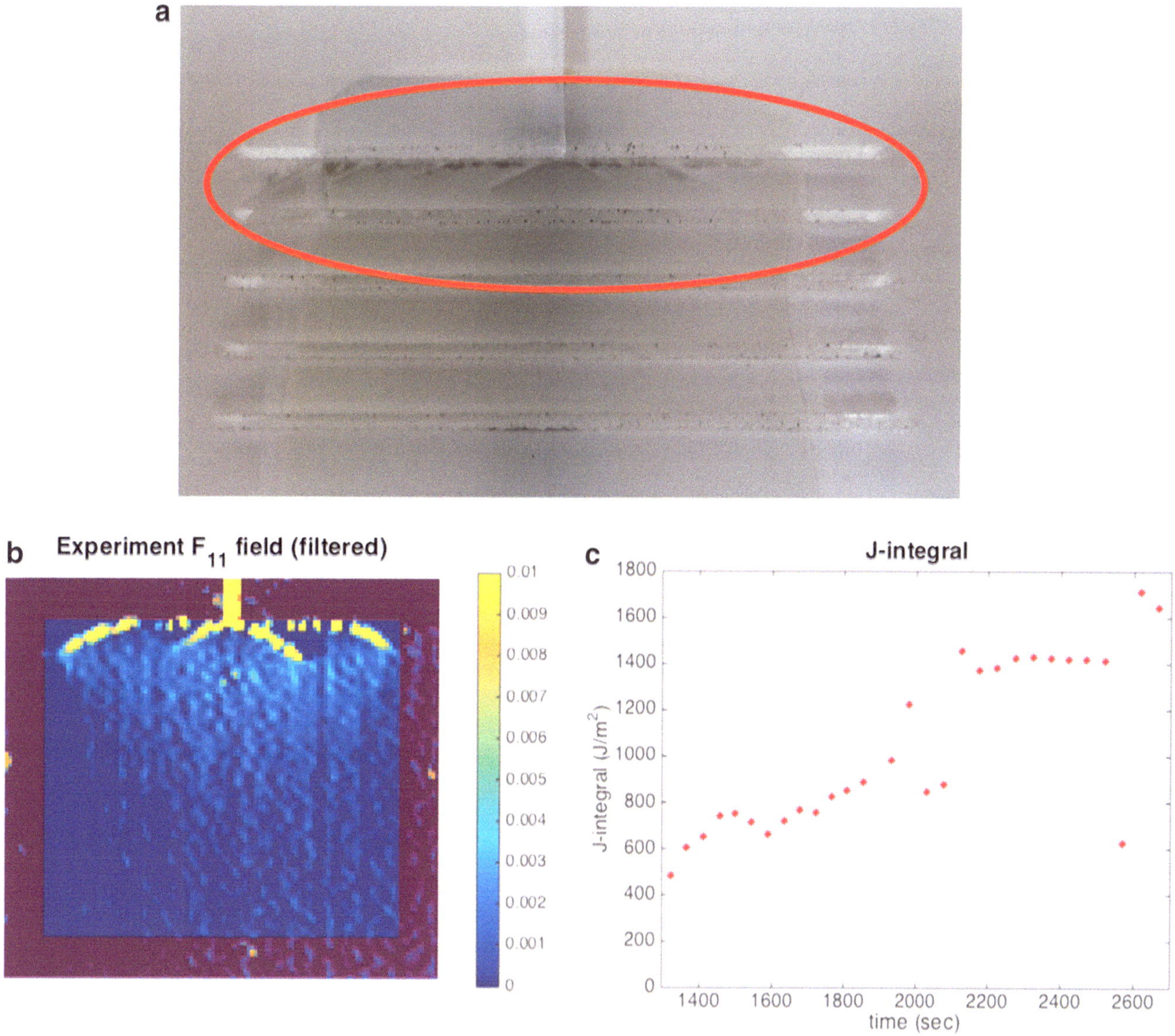


**Fig. 3.6** (**a**) Photograph of bifurcated crack after being trapped at the weak plane, (**b**) deformation gradient map and integration area for $J$-integral, (**c**) $J$-integral as a function of time during crack propagation

## 3.6 Conclusions

In this study, we proposed a new experimental configuration that gives steady crack growth macroscopically. This configuration enabled us to investigate crack propagation in heterogeneous materials. With this, we have measured the toughness of homogeneous and heterogeneous PMMA specimens. We saw that the toughness of homogeneous PMMA remained constant during the crack propagation, in agreement with values from the literature. We also found that the heterogeneity lead to toughening mechanisms and a substantial increase in the energy release rate. These results motivate are continued investigation into the interplay of heterogeneity and fracture as the experiments suggest that an improved understanding of this interplay could provide a route for engineering tougher materials through heterogeneity.

**Acknowledgement** We gratefully acknowledge the financial support of the National Science Foundation Award No. DMS-1535083 under the Designing Materials to Revolutionize and Engineer our Future (DMREF) program.

## References


1. Kruzic, J.J., et al.: The utility of R—curves for understanding fracture toughness—strength relations in bridging ceramics. J. Am. Ceram. Soc. **91**(6), 1986–1994 (2008)
2. Liang, F., Mayur, K.: Fracture behavior of flock fiber reinforced laminar composite. J. Mater. Sci. Eng. **1**(1B), 1 (2011)
3. Hossain, M.Z., et al.: Effective toughness of heterogeneous media. J. Mech. Phys. Solids **71**, 15–32 (2014)
4. Dupré, J.-C., Brémand, F., Lagarde, A.: Numerical spectral analysis of a grid: application to strain measurements. Opt. Lasers Eng. **18**(3), 159–172 (1993)
5. Surrel, Y.: Moiré and grid methods: a signal-processing approach. In: International Conference on Interferometry'94. International Society for Optics and Photonics (1994)
6. Grédiac, M., Pierron, F., Surrel, Y.: Novel procedure for complete in-plane composite characterization using a single T-shaped specimen. Exp. Mech. **39**(2), 142–149 (1999)
7. Badulescu, C., et al.: A procedure for accurate one-dimensional strain measurement using the grid method. Exp. Mech. **49**(6), 841–854 (2009)
8. Piro, J.-.L., Grédiac, M.: Producing and transferring low-spatial-frequency grids for measuring displacement fields with moiré and grid methods. Exp. Tech. **28**(4), 23–26 (2004)
9. Avril, S., et al.: Overview of identification methods of mechanical parameters based on full-field measurements. Exp. Mech. **48**(4), 381–402 (2008)
10. Avril, S., et al.: A full-field optical method for the experimental analysis of reinforced concrete beams repaired with composites. Compos. A: Appl. Sci. Manuf. **35**(7), 873–884 (2004)
11. Avril, S., Vautrin, A., Surrel, Y.: Grid method: application to the characterization of cracks. Exp. Mech. **44**(1), 37–43 (2004)
12. Badulescu, C., Grédiac, M., Mathias, J.D.: Investigation of the grid method for accurate in-plane strain measurement. Meas. Sci. Technol. **20** (9), 095102 (2009)
13. Grédiac, M., et al.: The virtual fields method for extracting constitutive parameters from full-field measurements: a review. Strain **42**(4), 233–253 (2006)
14. Rice, J.R.: A path independent integral and the approximate analysis of strain concentration by notches and cracks. J. Appl. Mech. **35**(2), 379–386 (1968)
15. Li, F.Z., Shih, C.F., Needleman, A.: A comparison of methods for calculating energy release rates. Eng. Fract. Mech. **21**(2), 405–421 (1985)
16. Atkins, A.G., Lee, C.S., Caddell, R.M.: Time-temperature dependent fracture toughness of PMMA. J. Mater. Sci. **10**(8), 1381–1393 (1975)

# Chapter 4
# Improved Hybrid Specimen for Vibration Bending Fatigue

**Onome Scott-Emuakpor, Tommy George, Casey Holycross, and Charles Cross**

**Abstract** A hybrid specimen was developed to minimize material used when assessing bending fatigue behavior with a vibration-based experimental procedure. Motivation for this work comes from the eagerness to quantify critical three-dimensionally printed gas turbine engine components at low cost under representative operating and stress conditions. The original vibration-based fatigue specimen (a whole, square plate) is capable of providing airfoil representative data at a lower cost than failing a fully manufactured airfoil. The reduced cost, though significant, was further reduced with the hybrid specimen. The most recently published iteration of the hybrid specimen produced an insert-plate system that required 95 % less material than the original specimen in order to gather fatigue data, and the hybrid specimen data compared favorably within a 95 % prediction interval of the original for Aluminum 6061-T6. Despite the successful comparison, improvements to the hybrid plate specimen were still necessary for accurately assessing fatigue behavior of more commonly used aerospace alloys. Specifically, improving repeatability in the experimental response, increasing the hybrid specimen excitability (i.e. reduce specimen system damping), and minimizing damage accumulation on the carrier plate of the hybrid specimen were critical to the efficiency of the characterized bending fatigue behavior. The proposed investigation addressed these issues for Titanium 6Al-4V, resulting in a more repeatable experimental procedure and results that agree with whole plate data.

**Keywords** Fatigue • Vibration • Titanium • HCF • Turbine engine

## Nomenclature

| | |
|---|---|
| $A$ | Stress-life regression parameter |
| $B$ | Stress-life regression intercept |
| $g$ | Gravitational constant (9.81 m/s$^2$) |
| $N_{fail}$ | Number of cycles to failure |
| $N_{step}$ | Number of cycles per step |
| $\mu\varepsilon_a$ | Microstrain amplitude |
| $\sigma_a$ | Stress amplitude |
| $\sigma_{fail}$ | Stress amplitude during failed step |
| $\sigma_{pr}$ | Stress amplitude during previous step |

## 4.1 Introduction

The romanticized concept of three-dimensionally printed (or additively manufactured) turbomachinery components has spawned an array of technical challenges for gas turbine engine designers. Specifically, validating material properties germane to component geometric features (thickness, surface roughness) and operating environments can be extensive when addressing the many uncertainties associated with metal-based additive manufacturing (AM) processes. Assessing physical properties of AM components can be conducted two ways: manufacture the exact component of interest and subject it to expected operational conditions; or develop geometrically representative test specimens capable of being studied under the same stress and environmental states as the component of interest. Based on the significant reduction in cost when compared

O. Scott-Emuakpor (✉) • T. George • C. Holycross • C. Cross
Aerospace Systems Directorate, Air Force Research Laboratory, Bldg 18D, Wright-Patterson AFB, OH 45433, USA
e-mail: onome.scott-emuakpor.1@us.af.mil

 
A.T. Zehnder et al. (eds.), *Fracture, Fatigue, Failure and Damage Evolution, Volume 8*,
Conference Proceedings of the Society for Experimental Mechanics Series, DOI 10.1007/978-3-319-42195-7_4

to actual component experimentation, developing a representative test specimen for engine componentry is desirous for High Cycle Fatigue (HCF) assessment.

HCF is one of the steepest obstacles faced by gas turbine engine designers aiming for gains on performance via component weight reduction. Aware of the possible horrors HCF could introduce to engines during operation, the Air Force, by means of the Engine Structural Integrity Program (ENSIP), suggests that all critical materials be HCF characterized for $10^9$ loading cycles if the material does not traditionally exhibit the phenomenon of fatigue endurance limit [1]. The challenge in characterizing HCF to $10^9$ cycles (gigacycle) is that current experimental methods are limited to relatively low operating frequencies. For example, using a conventional servohydraulic load frame, fatigue data is acquired at a rate of 40 Hz or, less conventionally, up to 1000 Hz for high actuation rate machines with load capabilities not suitable for metallic materials with standard geometries [2–4]. Though servohydraulic load frames are beneficial for acquiring a wide range of data under combined steady and amplitude load conditions, the duration required to reach a gigacycle is about 290 days of non-stop testing for a conventional machine. Furthermore, servohydraulic load frames more commonly subject specimens to axial tension-compression fatigue loads which do not correlate to the stress state that critical components such as airfoils could experience. To address representative frequencies and stress states likely associated with airfoils, a vibration-based bending testing methodology operating at high frequencies (>1200 Hz) was developed [4–6]. By applying a base excitation to a cantilevered plate specimen, the vibration-based method is capable of acquiring either uniaxial or biaxial bending fatigue life results at loading conditions similar to those that could cause HCF failure in turbine engine blades and vanes. This method has been successful in characterizing HCF of AM materials and comparing to cold-rolled results [7, 8]; however, since the method only requires crack initiation to define an HCF failure point, the bulk of the plate is not necessary for HCF characterization. This need to reduce the unacknowledged fatigue affected material during HCF characterization led to the development of a two-part system as a vibration-based bending test specimen. The hybrid specimen, also known as hybrid insert-plate system or Bruns-Zearley plate, is an inserted specimen fastened to a cantilevered, U-shaped carrier plate. The new vibration-based bending specimen is a successful alteration to the original whole plate, capable of characterizing HCF of Aluminum (Al) 6061-T6 using only 5 % of the original test specimen volume [9].

The work in this manuscript focuses on improving to the hybrid specimen and controlling the experimental procedure of the vibration-based method. For improving the specimen, focus is placed on reducing damage accumulation on the carrier plate portion and increasing the system excitability (i.e. reducing specimen system damping) during testing by redesigning the insert geometry. Repeatability of test response with respect to excitation and fatigue life variation between tests are key focuses when understanding and controlling experimental procedures such as plate and insert torque values, strain gage type, laser vibrometer and accelerometer locations, and surface roughness associated with polishing technique. The fatigue tests are conducted on Titanium (Ti) 6Al-4V, and the results of the improved hybrid specimen successfully compare to whole plate fatigue behavior between $10^5$ and $2 \times 10^6$ cycles.

## 4.2 Vibration-Based Bending Fatigue Theory

Stated in the Introduction Section, the vibration-based bending method was developed to experimentally acquire either uniaxial or biaxial fatigue life results at loading conditions similar to those that could cause HCF failure in turbine engine airfoils. The test is carried out by supplying a sinusoidal base excitation from an electrodynamic shaker to a cantilevered specimen and holding the excitation at a resonant frequency. In the case of uniaxial bending fatigue, the specimen is a 3.2 mm thick rectangular plate with 114.3 mm square test-sectional area (Fig. 4.1); this is the specimen geometry that was modified to a hybrid insert-plate system. The resonant frequency required to generate uniaxial bending loads is based on material composition and specimen dimensions. The dimensions are determined using a geometric design procedure which investigated vibrating plate physics [10, 11] for a maximum von Mises stress amplitude (the fatigue zone) away from the clamped-edge of the specimen [4, 5]. For the square dimensions of the uniaxial specimen, attaining maximum stress away from the clamped edge is achieved by exciting the specimen at a chord-wise bending (or two-stripe) mode, which is roughly 1600 Hz for most aerospace alloys with the geometry of Fig. 4.1. Figure 4.2 illustrates the vibratory analysis of the chord-wise bending mode. The loading behavior of the vibration-based specimen geometry in Fig. 4.2 is more representative of the HCF conditions a turbine engine airfoil could experience during operation than axial mechanical fatigue loading.

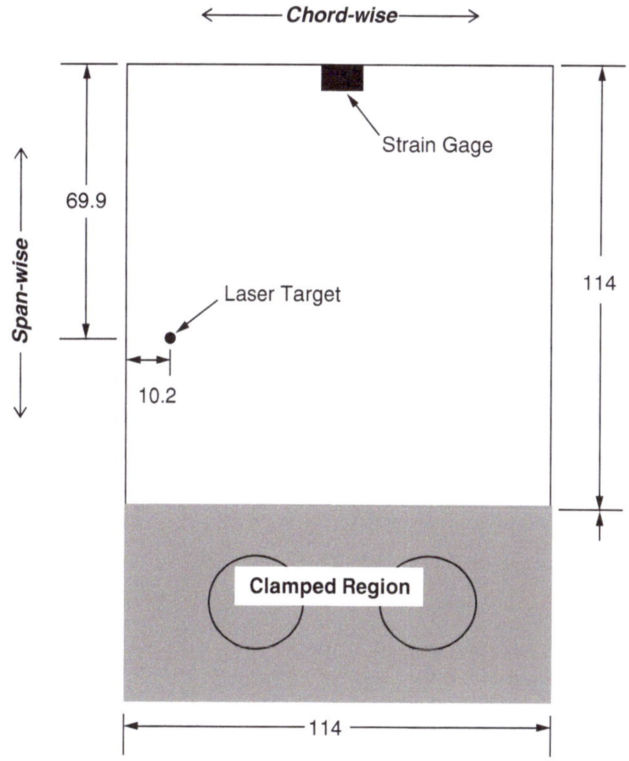


**Fig. 4.1** Vibration plate dimension (mm) and instrumentation locations [9]

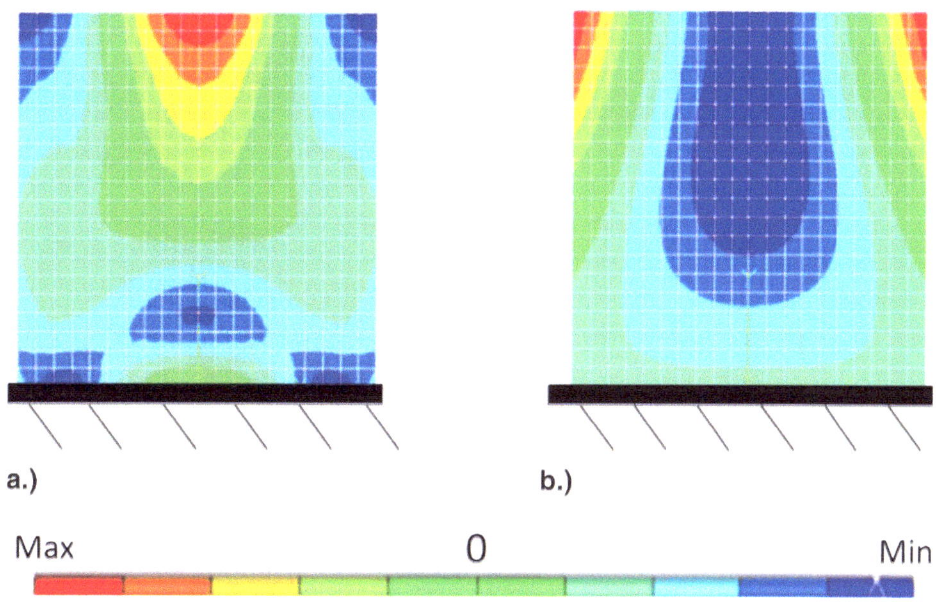


**Fig. 4.2** Vibratory analysis of the chord-wise mode on a square plate: (**a**) von Mises stress, (**b**) mode shape [7, 8]

## 4.3 Insert Specimen Redesign

The original hybrid specimen is illustrated in Fig. 4.3. Though the original insert geometry (rectangular) yielded useful fatigue results for Al 6061-T6, there were some occurrences of non-fatigue zone failures at either the insert-plate interface or under the reinforcing tabs [9]. Material imperfections were determined to be the cause of the premature failure; nonetheless, the results indicate a closer than anticipated vibratory strain amplitude relationship between the fatigue zone and other regions along the insert. For Ti 6Al-4V, material imperfections or poor surface treatment has a more adverse effect on

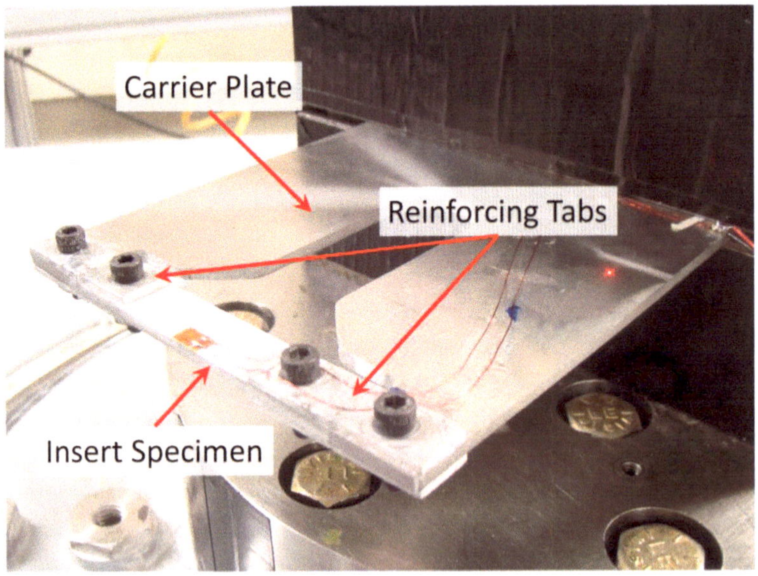


**Fig. 4.3** Original hybrid insert-plate system

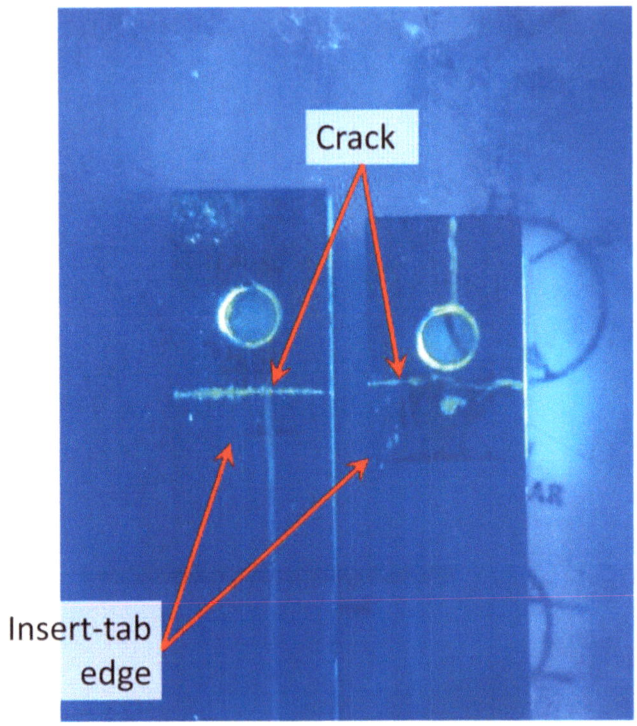


**Fig. 4.4** Fault penetrant inspection of Ti 6Al-4V rectangular inserts

fatigue behavior than Al 6061-T6 [12]. Therefore, for Ti 6Al-4V, the regions of the insert subjected to both fiction and alternating stresses (e.g. underneath the reinforcing tabs) could experience failure instead of the intended fatigue zone: a phenomenon also seen in mechanically fatigued axial dogbone specimens with grip-to-gage section ratios lower than two. This phenomenon was demonstrated in the first three tests of the hybrid specimen with rectangular Ti 6Al-4V inserts. The failure locations of two inserts are captured by Fig. 4.4. The picture shows that the failures occurred underneath the reinforcing tabs; the cracks are located 4 and 6 mm from the insert-tab edge. A more quantifiable example of surface preparation effects on fatigue results is shown later in the Experimental Results Section.

Along with the inability to generate a crack in the fatigue zone of the rectangular insert specimen, the ability to excite the hybrid system is of concern. For aerospace alloys that are stronger than Al 6061-T6, e.g. Ti 6Al-4V, the base acceleration required to generate HCF strain amplitudes challenges the threshold of most commercially available excitation systems. A solution to both this as well as the previously mentioned stress amplitude relationship issue can be achieved by

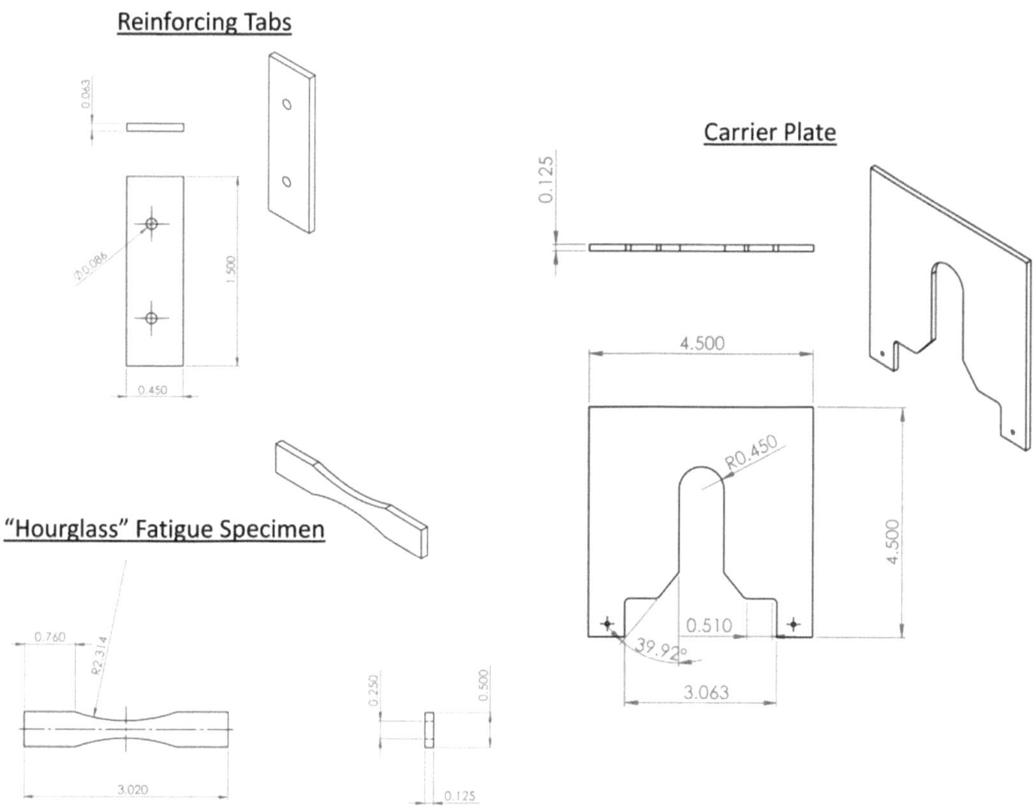


**Fig. 4.5** Hybrid insert-plate dimensions: insert, plate, and tabs

implementing a more compliant insert specimen than the rectangular geometry to the hybrid system. By leveraging the forced-controlled axial specimen design recommendations in ASTM E 466 [2], a flat hourglass (dogbone) insert with higher fatigue zone strain amplitude compared to the insert-plate interface, verified via finite element forced-response analysis, has been created. The dimensions for the hourglass specimen as well as other key components of the improved hybrid specimen are illustrated in Fig. 4.5.

## 4.4 Experimental Procedure

Unlike the single boundary condition (BC) of the original vibration-based bending experiments, the hybrid specimen essentially has three BC: a BC within a BC within the original cantilever, or clamped BC (see Fig. 4.6a). This highly coupled setup can create uncertainty in the specimen response capability with respect to excitation of the driving acceleration. The main BC, the cantilevered connection between the carrier plate and a steel disk mounted atop the excitation source, was solidified for both alignment purposes (ensuring symmetric 114.3 mm span) and clamping pressure with an early fixture redesign [13]. As for the BCs coupling the insert and plate, there is no built-in assurance for maintaining repeatability in the alignment of the connection. For this reason two steps were introduced to the fastening procedure of the hybrid specimen: torque limit and sequence. The 8–32 bolts (English unit classification) connecting the insert to the plate are torque rated at 3.4 m-N but were empirically verified with four destructive torque tests to a limit of 6.2 m-N. The empirically verified torque rating (6.2 m-N) is used for every hybrid specimen assembly. The nut-bolt connection is secured with 242 Threadlocker Locktite. The fastening sequence for the insert-tab and tab-plate BCs are, first, torque the inside bolts (insert-tab BC), and then torque the outside bolts (tab-plate BC). The clamped BC torque value is set at 170 m-N for both the inner (1/2 in.) and the outer (3/8 in.) studs.

The general procedure for testing the hybrid specimen is similar to the whole plate vibration-based fatigue test [4, 8, 9]. Base excitation for the test is produced by a 90 kN Unholtz-Dickie electrodynamic shaker. Instrumentations used to measure data during testing are a PCB Piezotronics 352B10 accelerometer, a Micro-Measurements CEA-05-062UW-350 (titanium specific) uniaxial strain gage, and a Polytec OFV 505 single-point laser vibrometer for non-contact data. The accelerometer is used to

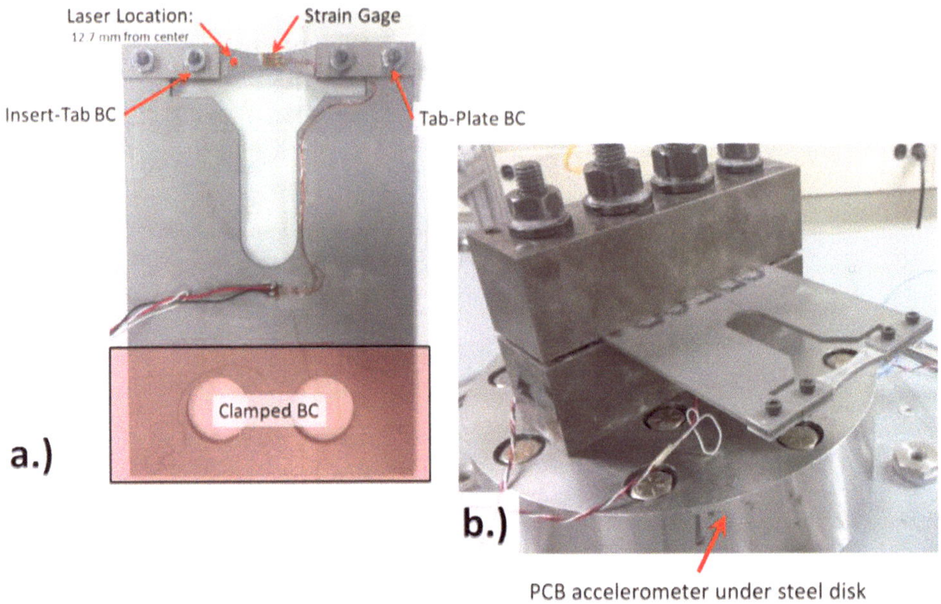


**Fig. 4.6** (**a**) Assembled hybrid specimen with instrumentation. (**b**) Hybrid specimen clamped on shaker

control the acceleration force of the shaker armature, the strain gage is placed at the fatigue zone of the insert to acquire strain amplitude, and the laser point of the vibrometer is positioned on the insert near the system nodal line (0.25″ from the insert-plate edge) to reduce the likelihood of velocity measurement saturation (see Fig 4.6 for instrumentation locations) [4, 9]. The strain gage used for fatigue testing are not rated for the strain amplitudes and accumulated cycles investigated in this study; gages are rated for less than 1500 $\mu\varepsilon_a$ and $10^5$ cycles. Therefore, in order to determine Ti 6Al-4V HCF appropriate strain amplitude values (3500–6000 $\mu\varepsilon_a$) throughout the duration of a test ($10^5$-2 × $10^6$ cycles), the velocity results from the laser vibrometer are linearly correlated with strain results from the gage at sufficiently low strain amplitudes (below 1500 $\mu\varepsilon_a$) and low loading cycles (15 Hz sinusoidal sweeps around resonance for 180 s durations). The velocity response is then used to identify strain/stress amplitudes during fatigue testing.

Failure is determined during testing by monitoring frequency change with respect to time [14]. Also, observation of the shaker acceleration force versus time is monitored to support the frequency change. Failure is identified as the instance the natural frequency, which is tracked by the driving frequency, shifts more than 0.1 % [9, 15]. This frequency shift is due to a change in the stiffness of the plate as a crack appears, and can be associated with crack lengths as small as 2–3 mm [4, 15]. Damping in the plate also increases with the presence of a crack, thus more force (i.e. shaker head acceleration) is required to maintain a desired velocity response.

Fatigue testing to failure is conducted by the step test procedure. This procedure is a series of sinusoidal amplitude dwell steps for a specified number of cycles ($N_{step}$), each step having consecutively increasing sinusoidal dwell amplitudes until failure occurs. The failure stress amplitude is calculated by interpolating between the failure and previous steps [16, 17]. Equation (4.1) represents the failure stress amplitude ($\sigma_a$) of $N_{step}$ cycles to failure. This procedure has been successfully validated against constant sinusoidal dwell results for aluminum, titanium, and nickel-based alloys [7, 18–20].

$$\sigma_a = \sigma_{pr} + \frac{N_{fail}}{N_{step}}\left(\sigma_{fail} - \sigma_{pr}\right) \tag{4.1}$$

## 4.5 Experimental Results

Vibration-based bending fatigue behavior was characterized from 3.2 mm thick colled-rolled Ti 6Al-4V with the chemical composition listed in Table 4.1. Nine hourglass inserts along with carrier plates were waterjet cut from the same Ti 6Al-4V plate stock. The surfaces of the inserts were hand prepared to a smooth finish (<0.13 μm Ra) with a two-step sanding process of 120 and 600 grit paper. The carrier plate surfaces were not polished or treated after waterjet cut. Only one carrier plate with the same braided gage wire and lead tabs was used to gather hybrid specimen fatigue data. The integrity of the carrier plate along with the associated components is discussed later in this section.

**Table 4.1** Ti 6Al-4V chemical composition

| Ti | Al | C | Si | Mn | Mo | V | Fe | Cu |
|---|---|---|---|---|---|---|---|---|
| 89.02 | 6.62 | 0.016 | <0.01 | <0.01 | <0.01 | 4.16 | 0.16 | <0.01 |

**Table 4.2** Tensile properties of whole and hybrid plate Ti 6Al-4V stocks

| Specimen | E (GPa) | Yield (MPa) |
|---|---|---|
| Whole | 117.5 | 1037 |
| Hybrid | 119.7 | 1014 |

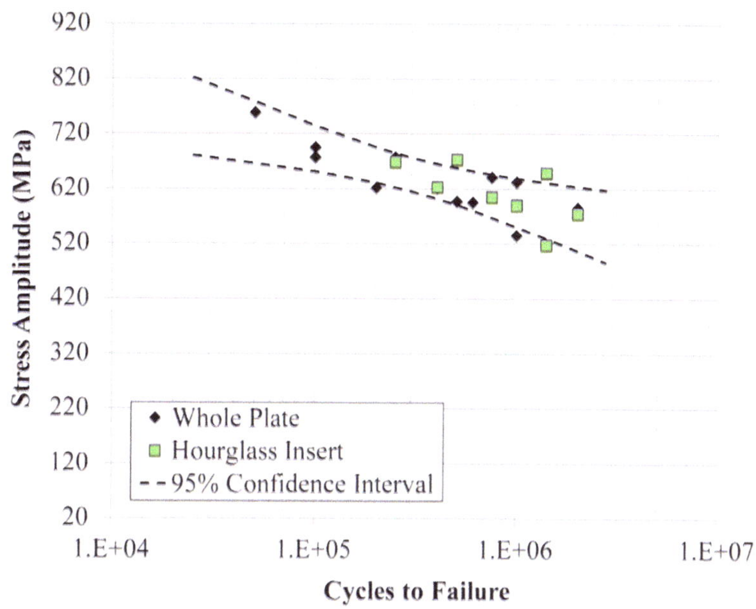


**Fig. 4.7** Ti 6Al-4V vibration bending fatigue life comparison

Fatigue behavior of the hybrid specimen was compared against whole plate (original vibration bending specimen) data from a different cold-rolled Ti 6Al-4V stock [14]. The average empirical modulus and tensile yield strength for the hybrid and whole plate stocks were determined via standard servohydraulic tensile testing [21] and are listed in Table 4.2 to assure similarities in the bulk materials. Fatigue behavior comparison is plotted in Fig. 4.7. Fatigue data was gathered from only eight of the nine inserted specimens due to instrumentation uncertainty during one of the sinusoidal amplitude dwell processes. For the hybrid specimen, there is a large fatigue stress amplitude discrepancy at $1.4 \times 10^6$ cycles. Closer observation of this difference probed deeper into the surface finishing process. The specimen used to acquire the lower stress amplitude (516 MPa) data point had a maximum Ra value (surface roughness parameter) of 6.5 μm; the roughness is representative of a post-waterjet finish. The specimen with the higher stress amplitude had a maximum Ra value of 0.1 μm that was achieved by using two types of polishing applications: pneumatic air polisher and the two-step sanding procedure mentioned earlier in this Section. The higher stress amplitude of 648 MPa at $1.4 \times 10^6$ cycles to failure is a relatively high value for Ti 6Al-4V. The high stress value may be an artifact of the pneumatic air polisher unintentionally adding compressive residual stresses that credited the life of the inserted specimen [22]. As for the remaining hybrid specimen data, i.e. ignoring the outliers at $10^6$ cycles for the whole plate specimen and $1.4 \times 10^6$ cycles for the hybrid specimen, the comparison between the two specimens shows more consistency than the Al 6061-T6 comparison of the former hybrid specimen design [9] as the data fits within a 95 % confidence interval (Fig. 4.7) [23, 24]. Further observation of the data after omitting the outliers shows the correlation coefficient between the whole plate regression line (represented by Eq. (4.2)) and the hybrid data is R = 0.812 [23]. The correlation coefficient of hybrid data compared to whole plate data for the same regression line (R = 0.831) also supports the consistency of the two fatigue data sets and shows a strong linear relationship [24] for both stress amplitude and cycles to failure data sets.

$$\sigma_a = A\log\left(N_{step}\right) + B \tag{4.2}$$

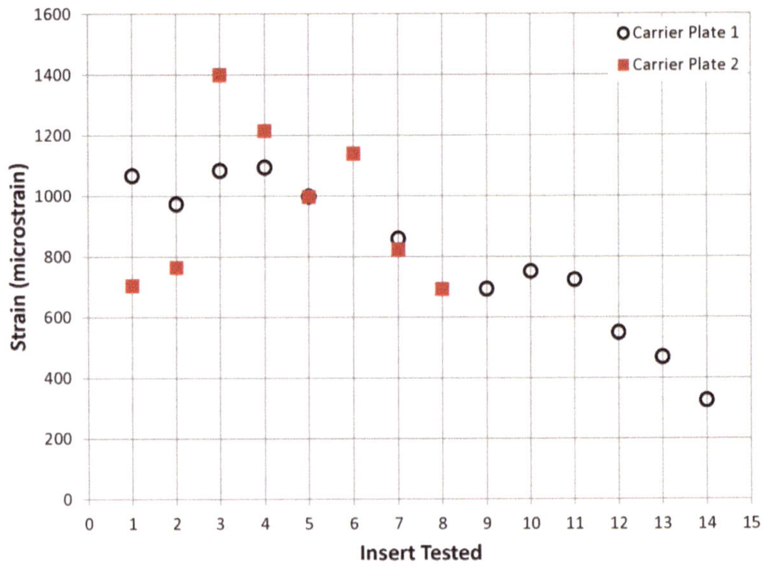


**Fig. 4.8** Al 6061-T6 carrier plate strain amplitude response for inserts at 10 g base acceleration [9]

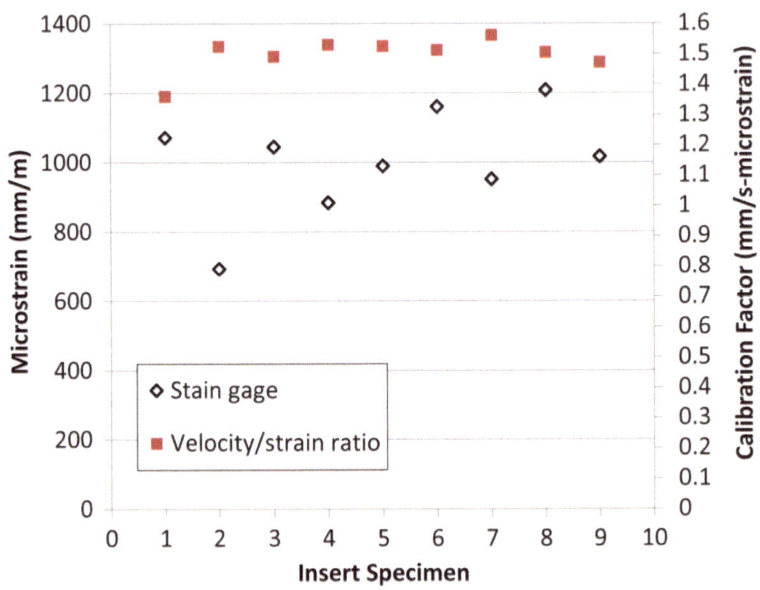


**Fig. 4.9** Ti 6Al-4V carrier plate strain amplitude response for insert at 10 g base acceleration

Previous work on the hybrid specimen showed degradation in the strain response versus shaker excitation presumably due to carrier plate damage [9]. This degradation is illustrated on the plot of Fig. 4.8 for the Al 6061-T6 rectangular inserts. The same observation was made with the single Ti 6Al-4V carrier plate used to gather the fatigue data in Fig. 4.7. These response results, recorded at 10 g base acceleration from the shaker, are plotted in Fig. 4.9. Also plotted in Fig. 4.9 is the laser-to-strain response calibration (or correlation) factor mentioned in the Experimental Procedure Section. The strain response plotted in Fig. 4.9 shows comparable variation to the response of Al 6061-T6 in Fig. 4.8. Unlike Fig. 4.8, however, the Ti 6Al-4V strain response does not appear to exhibit a decline with respect to base acceleration like the Al 6061-T6 case. Also, aside from the first inserted Ti 6Al-4V specimen, the calibration factor for each hybrid specimen test shows consistency within a relative standard deviation of 1.8 %. The gage wire survived the duration of nine fatigue tests. The lead tabs, however, damaged during a sinusoidal dwell step of the ninth tested insert. As a whole, the carrier plate, gage wire, and lead tabs maintained integrity for nine tests with over 30 million combined cycles at sinusoidal amplitude greater than 3500 $\mu\varepsilon_a$.

## 4.6 Conclusion

The hybrid insert-plate specimen that was developed to minimize the amount of material required to assess vibration bending fatigue behavior has been improved by defining a repeatable experimental procedure and redesigning the insert geometry of the system from rectangular to hourglass. The new specimen and experimental procedure for the hybrid specimen was validated with a comparison against whole plate specimens which used 95 % more unnecessary material for fatigue characterization; both the hybrid and whole specimens were manufactured from cold-rolled Ti 6Al-4V sheets.

The fatigue results showed similar linear comparisons for the whole and hybrid life data via correlation coefficient comparisons as well as a 95 % confidence interval. Furthermore, unlike the results seen in the strain response study conducted on Al 6061-T6, the carrier plate showed no signs of degradation due to repeated use for over 30 million combined cycles at system sinusoidal amplitudes greater than 3500 $\mu\varepsilon_a$. The nine tests conducted also showed consistency in the velocity-to-strain ratio calibration factor within a relative standard deviation of 1.8 %.

The improvements to the vibration-based test procedure of a hybrid bending specimen gives confidence in the future assessment of fatigue behavior for high strength aerospace alloys. This procedure will be used in the near future for the assessment of specimens made via directed energy deposition and laser powder bed fusion additive manufacturing processes.

**Acknowledgements** The authors would like to thank the following organizations for funding and collaboration: the Turbine Engine Fatigue Facility (TEFF) of the United States Air Force Research Laboratory (AFRL) and Universal Technology Corporation (UTC). Specifically, the authors would like to acknowledge UTC contractors Phil Johnson, Jeff Bruns, and Alyssa Zearley for their contribution to specimen and fixture design and analysis, as well as laboratory testing and experimentation setup.

## References

1. Department of Defense Handbook: Engine Structural Integrity Program (ENSIP), MIL HDBK-1783B (USAF), 15 Feb 2002
2. American Society for Testing and Materials: E466-07: standard practice for conducting force controlled constant amplitude axial fatigue tests of metallic materials. In: ASTM Book of Standards, vol. 03.01. ASTM International, West Conshohocken, PA (2009)
3. Scott-Emuakpor, O., Shen, M.-H.H., George, T., Cross, C., Calcaterra, J.: Development of an improved high cycle fatigue criterion. ASME J. Eng. Gas Turbines Power **129**(1), 162–169 (2007)
4. George, T., Seidt, J., Shen, M.-H.H., Cross, C., Nicholas, T.: Development of a novel vibration-based fatigue testing methodology. Int. J. Fatigue **26**, 477–486 (2004)
5. George, T., Shen, M.-H.H., Cross, C., Nicholas, T.: A new multiaxial fatigue testing method for variable amplitude loading and stress ratio. ASME J. Eng. Gas Turbines Power **128**, 857–864 (2006)
6. George, T., Shen, M.-H.H., Scott-Emuakpor, O., Nicholas, T., Cross, C., Calcaterra, J.: Goodman diagram via vibration-based fatigue testing. ASME J. Eng. Mater. Technol. **127**(1), 58–64 (2005)
7. Scott-Emuakpor, O., Schwartz, J., George, T., Holycross, C., Cross, C., Slater, J.: Bending fatigue life characterization of direct metal laser sintering nickel alloy 718. Fatigue Fract. Eng. Mater. Struct. **38**(9), 1105–1117 (2015)
8. Scott-Emuakpor, O., Holycross, C., George, T., Knapp, K., Beck, J.: Fatigue and strength studies of titanium 6Al-4V fabricated by direct metal laser sintering. ASME J. Eng. Gas Turbines Power **138**(2) (2015), paper No. GTP-15-1306
9. Bruns, J., Zearley, A., George, T., Scott-Emuakpor, O., Holycross, C.: Vibration-based bending fatigue of a hybrid insert-plate system. J. Exp. Mech. **55**(6), 1067–1080 (2015)
10. Simons, D., Leissa, A.: Vibrations of rectangular cantilever plates to in-plane acceleration loads. J. Sound Vib. **17**(3), 407–422 (1971)
11. Meirovitch, L.: Principles and Techniques of Vibrations. Prentice Hall, New Jersey (1997)
12. Mower, T.: Degradation of titanium 6Al-4V fatigue strength due to electrical discharge machining. Int. J. Fatigue **64**, 84–96 (2014)
13. Blackwell, C., Palazotto, A., George, T., Cross, C.: The evaluation of the damping characteristics of a hard coating on titanium. Shock. Vib. **14**, 37–51 (2007)
14. Scott-Emuakpor, O., Shen, M.-H.H., George, T., Cross, C.: A energy-based uniaxial fatigue life prediction method for commonly used gas turbine engine materials. ASME J. Eng. Gas Turbines Power **130**(6) (2008), paper No. 062504
15. Bruns, J.: Fatigue crack growth behavior of structures subject to vibratory stresses. In: Society of Experimental Mechanics Annual Conference, International Student Paper Competition, Greenville, SC, 2–6 June (2014)
16. Maxwell, D.C., Nicholas, T., A rapid method for generation of a Haigh diagram for high cycle fatigue. J. Fatigue Fract. Mech., ASTM STP 1321, **29**, 626–641 (1998)
17. Nicholas, T.: High Cycle Fatigue: A Mechanics of Materials Perspective. Elsevier, Oxford (2006)
18. Scott-Emuakpor, O., Holycross, C., George, T., Beck, J., Schwartz, J., Shen, M.-H.H., Slater, J.: Material property determination of vibration fatigued DMLS and cold-rolled nickel alloys. ASME/Turbo Expo, Dusseldorf, Germany, 16–20 June 2014, Paper No. GT2014-26247
19. Bellows, R., Muju, S., Nicholas, T.: Validation of the step test method for generating Haigh diagrams for Ti-6Al-4V. Int. J. Fatigue **21**(7), 687–697 (1999)

20. Scott-Emuakpor, O., George, T., Cross, C., Wertz, J., Shen, M.-H.H.: A new distortion energy-based equivalent stress for multiaxial fatigue life prediction. Int. J. Non Linear Mech. **47**(3), 29–37 (2012)
21. American Society for Testing and Materials: E 8M-09: standard test methods for tension testing of metallic materials. In: ASTM Book of Standards, vol. 03.01. ASTM International, West Conshohocken (2009)
22. Prevey, P., Jayaraman, N., Ravindranath, R.: Introduction of residual stress to enhance fatigue performance in the initial design. ASME/Turbo Expo, Vienna, Austria, 14–17 June 2004, Paper No. GT2004-53971
23. American Society for Testing and Materials: E739-10: standard practice for statistical analysis of linear or linearized stress-life (S-N) and strain-life (ε-N) fatigue data. In: ASTM Book of Standards, vol. 03.01. ASTM International, West Conshohocken (2009)
24. Kutner, M., Nachtsheim, C., Neter, J.: Applied Linear Regression Models. McGraw-Hill/Irwin, New York (2004)

# Chapter 5
# Experimental Study of Residual Plastic Strain and Damages Development in Carbon Fiber Composite

**Addis Tessema, Suraj Ravindran, and Addis Kidane**

**Abstract** In this work, the correlation between local damage and residual stiffness in carbon fiber composites is investigated experimentally. The study presents the interaction between local damage mechanisms and the global response of the material, with the aim to capture the gradual local failure phenomenon. High resolution digital image correlation technique at micro length scale is used to measure the local deformation in the specimen made of carbon fiber composite, subjected to tension-tension fatigue loading. During testing, image of a speckled surface inside the gage length of the specimen is captured at specified number of loading cycles. Using digital image correlation, the acquired images are processed and the local deformations and strains are extracted. The growth of local plastic deformation as a function of loading cycle is acquired and different damage modes as a function of loading cycle is explored. The study has able to elucidate the event that shows the gradual growth of matrix cracking into inter-ply debonding. Furthermore the degradation of modulus of elasticity as a function of loading cycle is determined and the corresponding type of damage incorporated within the plastic strain distribution around it is presented.

**Keywords** Residual strain • Damage evolution • Matrix crack • Stiffness degradation • Fatigue loading

## 5.1 Introduction

Intensive studies have been conducted in the past, to investigate the causes, conditions and mechanism for the initiation and development of damages. There are many types of damages such as; transverse cracking, ply delamination, fiber splitting and fiber breaking, that are identified and reported in the literatures. Usually two or more type of damages can exist at the same time even though one form of damage could dominate over the other. The form of damage mainly depends on the type of loading scheme, thickness of laminate, the geometry of the structure, ply orientation and stacking sequence [1, 2]. Once damage is nucleated, the initiation and direction of propagation are directly influenced by the fiber angle arrangement of ply relative to the loading direction [1]. Through an optical microscope observation of failed specimens, it was observed that there is a transformation of damages from one form to another form.

Recently, using techniques like fiber architecting, the sequence and orientation of fibers/plies within the laminate are tailored to sustain the anticipated major loads in specific direction. For example fibers are aligned with the direction of the main axial loads (on-axis), while some other fibers are arranged at a certain angles (off-axis) to provide the necessary shear and transverse stiffness within a laminate. Among the different type of conventional ply arrangements cross-ply and quasi-isotropic are the most common.

Different theories and analytical models have been developed to elaborate the relation between damage type and density with material properties. These theories are based on classical laminate theory [2], fracture mechanics [3, 4], energy method [5] and constitutive/elasticity models [6] and are mainly focused understanding the effect of damage on the mechanical properties of the composites. An experimental based work, on understanding how these damages are initiated and evolved to macro level failure, is missing in the literatures.

Recently, digital image correlation technique has been applied in understanding material deformation at different length scale from macro to nano scale [7–10]. Most recently, DIC has been effectively used to get the strain localization in composites at meso [9, 11] and micro [12, 13] scales. These small scale strain localizations have been helpful in formulating a correlation between the microstructural element and its displacement with macro scale deformation. The main objective of this study is to evaluate the gradual evolution of damage in composites during fatigue loading using in-situ DIC technique at

A. Tessema (✉) • S. Ravindran • A. Kidane
Department of Mechanical Engineering, University of South Carolina, 300 Main Street, Columbia, SC 29208, USA
e-mail: atessema@email.sc.edu



A.T. Zehnder et al. (eds.), *Fracture, Fatigue, Failure and Damage Evolution, Volume 8*,
Conference Proceedings of the Society for Experimental Mechanics Series, DOI 10.1007/978-3-319-42195-7_5

high magnification. The initiation and propagation of micro cracks at different cycle are captured. Further the local strain distribution around the crack is assessed during loading, and the transformation of cracks from one form to the other is observed. Finally the influence of damages on the modulus of elasticity of the laminate is investigated.

## 5.2 Material Preparation

As manufactured unidirectional carbon Prepregs (TC350-1) purchased from Royal TenCate Cooperate are used to prepare quasi-isotropic and cross-ply laminates. The Prepregs are cut in square of 150 × 150 mm size at 0° and 45° fiber angle orientation using a scissors. The Prepreg cut outs are stacked in the required sequence and orientation $(\pm 45/90/0)_s$, and placed in the plastic bag and kept in between two aluminum mold plates. The mold is placed in a hot press machine, and clamped between the platens at a pressure of 3.5 ton and set to the curing temperatures recommended by the manufacturer. The curing cycle has three stages, first the machine platens are heated up to temperature of 225 °F at rate of 2 °F/min and kept for 1 h, then the platens temperature is heated to temperature of 350 °F at the same rate and kept for 2.5 h, and the machine is cooled to a temperature of 120 °F at the rate of 5 °F/min. Finally, the cured laminate is taken out of the hot press and let to cool down to room temperature naturally.

According to ASTM D3930 standard rectangular samples (140 mm × 22 mm) are cut from the cured laminate using CNC water jet. Then, the side surfaces are polished in order to remove any debris and notches formed during the abrasive cutting by the water jet. Fiberglass tabs are glued on the two ends of the specimen in order to protect the fibers breaking in the samples due to the grip pressure during the experiment. For digital image correlation, two different size speckle patterns are applied on the edge and the width of the samples. To capture the local strain on the plies, the edge of the samples is speckled with approximately 6 ~ 16 μm size black and white patterns. To obtain the laminate average strain, the width face of the samples is speckled with approximately 250 ~ 300 μm size black and white patterns.

## 5.3 Experimental Set-Up

Tension-Tension fatigue experiment is conducted based on ASTM D3479 standard, using Material Testing System (MTS 810) machine shown in Fig. 5.1. The machine is programmed to apply two kinds of loadings alternatively, monotonic quasi-static and fatigue loadings. The monotonic quasi-static loading is performed intermediately between certain fatigue cycles to determine the modulus of elasticity of the laminate after the predefined cycles. The quasi-static test is performed at the

**Fig. 5.1** (**a**) Experimental set-up. (**b**) Test sample with macro speckles. (**c**) Sample side view with micro speckles

loading rate of 0.01 mm/min and loaded up to stress of 200 MPa (i.e. 30 % of the failure stress). The fatigue test is executed with mean stress of 165 MPa and amplitude stress of 135 MPa at a frequency of 5 Hz.

The full field in-plane strains are measured using 2D digital image correlation (2D-DIC) for both speckled regions. For the micro speckle region, the area of interest is very small (i.e. 3 mm × 1 mm), a 3× magnification lens is mounted on a 9 M Pixels CCD (Grasshopper) camera to capture the images at an object distance of 20 mm. Images of the micro speckle region are taken at predefined cycles by stopping the fatigue and bringing the applied load to zero. On the other side, the images of the macro speckled region of the width face are acquired using 60 mm Nikon lenses incorporated with 9MPixel CCD cameras at a framing rate of 1 fps. The recorded images are processed using VIC-2D commercial software and the fulfield axial, transverse and shear strain data are obtained at local and far-field locations.

## 5.4 Results and Discussion

### *5.4.1 Damage Evolution and Modulus Degradation*

As shown in Fig. 5.2a, degradation in modulus of elasticity/stiffness with the loading cycle is observed in this study. It is presumed that the damage was initiated at the location where there is a defect or pores on the 90° plies. The damages are initiated as transverse cracks and expands gradually to the neighboring plies as shown Fig. 5.2b. The first matrix cracks within our area of interest are initiated around the first hundreds of cycles and at the same time a drop in modulus of elasticity is observed as depicted in Fig. 5.2a. As the cycle grows further, the number of newly formed matrix cracks in the 90° and 45° plies have increased. In parallel, the pre-formed matrix cracks are growing towards the adjacent interfaces. At this time, the modulus of elasticity has decreased by about 7 % when the loading cycles reached around 3000. It is also noticed that, when the crack reached at the interface of the adjacent ply, it is arrested for a while. This crack arrest could be due to the resistance

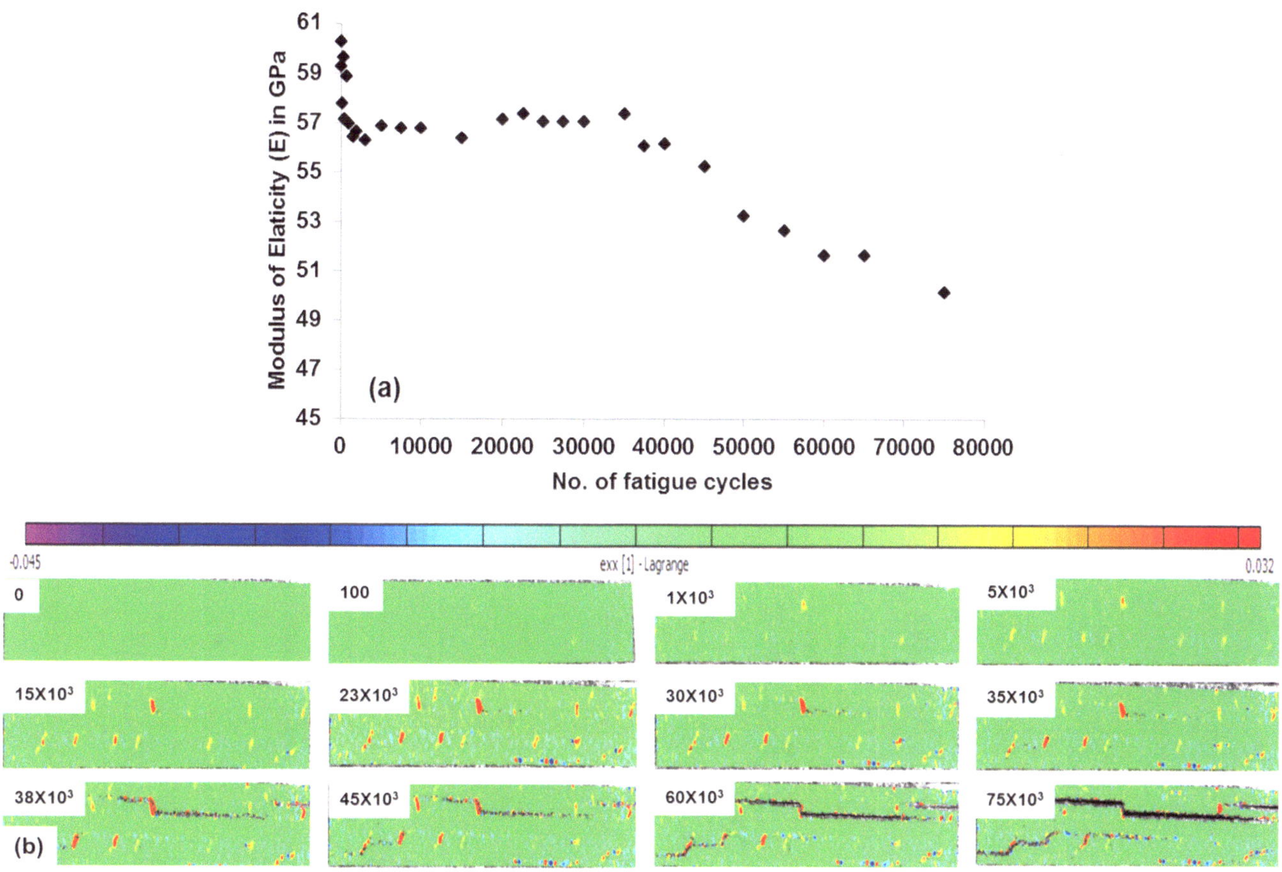


Fig. 5.2 (a) Variation of modulus of elasticity with fatigue cycle and (b) gradual development of micro damages and their local strain distribution

provided by the neighbor plies. Depending on the fiber orientation angle of the adjacent ply, the cracks either propagates to the next ply or the interface and cause delamination. As it is shown in the Fig. 5.2b, the cracks from 90° plies are shown to proceed to the next +45 ply. But, the crack which approached to the interface of the 0° ply is observed to stay at the interface and resulted delamination.

The rate of degradation in the elastic modulus has decreased for the cycles range between 5000 to 40,000. In this range, delaminating cracks are just initiated and the formation of new matrix cracks has decreased or saturated (known as Characteristic Damage State (CDS)), as a result, the rate of degradation in modulus is very slow. At higher loading cycles, most of the transverse/matrix cracks has transformed to delamination and these delaminating cracks have grown well, and started to merge into one continuous delamination crack that separates the off-axis plies from the on-axis plies.

### 5.4.2 *Crack Initiation and Gradual Development*

One of the primary matrix cracks formed within our area of interest is analyzed and the progressive change in axial and shear strains is plotted as shown in the Fig. 5.3. It is clearly depicted that both the axial and shear strains have the same trend along with the fatigue cycles, and it is apparent that during the early stage of the matrix crack both the axial and shear strain have increased sharply up to 5000 cycles. As the cycle advances, the strains seems to remain constant up to 15,000 cycles, and this could be due to the crack reaching the interface and arrested from going any further. Afterwards, the sudden uprising of strains is observed, and this is taken as an evidence for initiation of delamination. Earlier, it was observed that the source of the delaminating cracks are matrix cracks and resulted where they reached the interface. When the delamination is initiated, the surfaces of the matrix crack totally freed from the neighbor plies constraint, which can lead to a sudden contraction of the crack faces in the loading direction. As a result, a jump in the strain between the two crack faces is observed as shown in the figure.

### 5.4.3 *Damage Sequences and Evolution*

To investigate the evolution of different cracks as a function of loading cycle, the strain at different crack locations (the cracks along the dashed line) as shown in Fig. 5.4a) are compared. It was observed that, the axial and shear strain at the three crack points varied with loading cycle. It is clearly visible from Fig. 5.4b, c that the increase in strain at these three

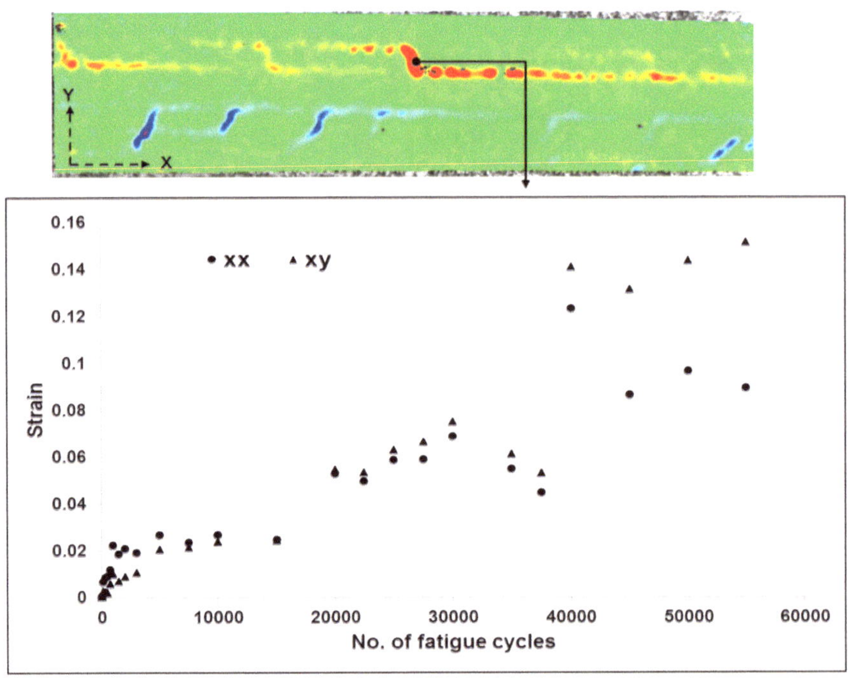


**Fig. 5.3** Axial and shear strain profiles for a point near the matrix crack with fatigue cycle

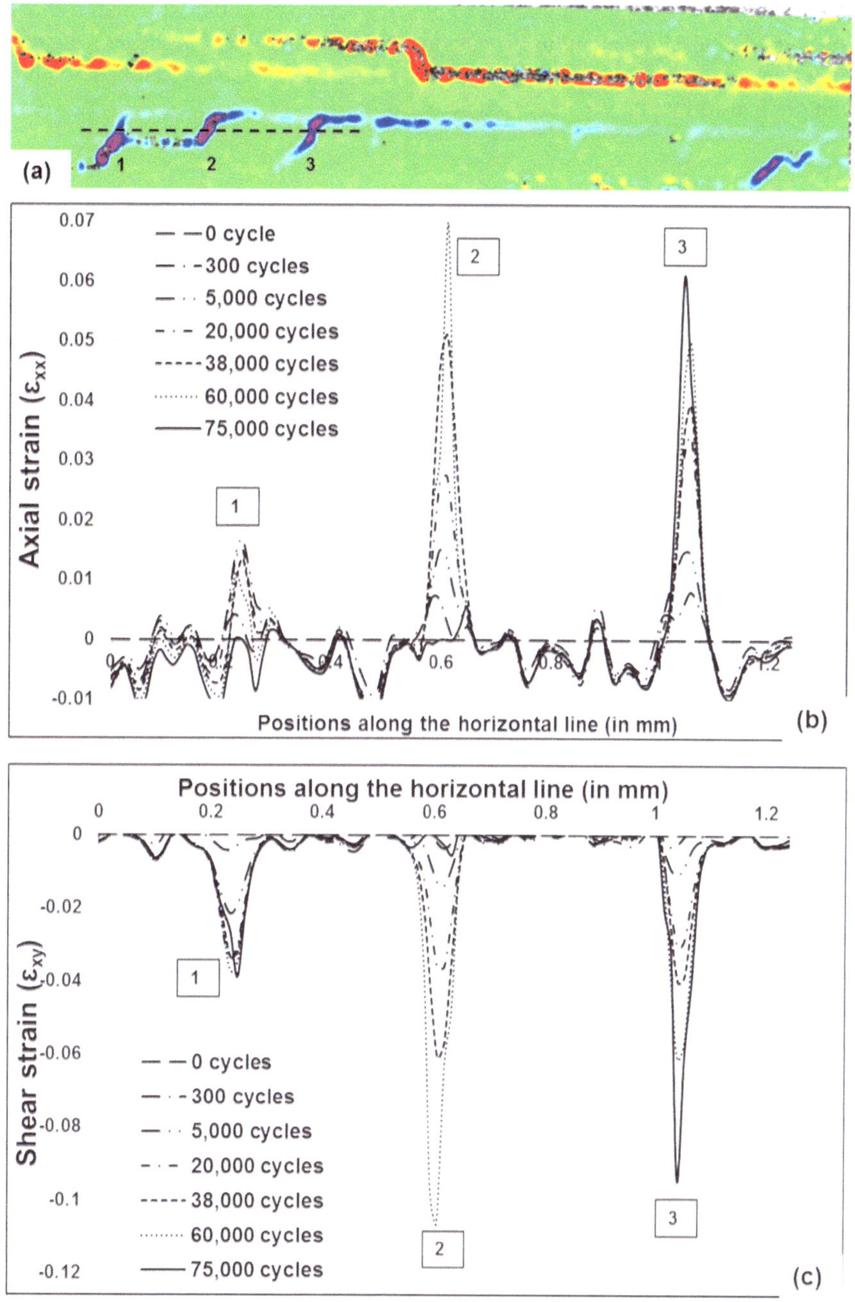


**Fig. 5.4** (**a**) Strain contour on the body. (**b**) Axial strain variation. (**c**) Shear strain variation along the position on the line

points are not the same, though the trend is similar. For example the strain at the middle crack (#2) reached the peak value way ahead compared with the other two cracks. Interestingly, the strain in the middle crack decreased later as the strain in the right crack (#3) reached its peak value. This can be taken as an indication that there is a progressive crack growth and time lagging between cracks along the same ply.

## 5.5 Conclusion

Using a high magnification an optical imaging system and digital image correlation, the local strain distribution in composites is analyzed. The progressive growth of damages within the quasi-isotropic laminate is captured along with degradation in modulus of elasticity during the fatigue loading. The gradual transformation of damage from transverse cracking to interface delamination is observed for the crack that grows from 90° ply to the 0°. The corresponding relation of damage with the change in modulus of elasticity has shown that, the matrix cracking and delamination has a loud effect on the degradation of the modulus. The local strain profile for a point near to the matrix crack has revealed the contraction and relaxation of crack faces during delamination. Finally, considering groups of matrix cracks exist in one region, the sequential and gradual formation of cracks are detected.

## References

1. Tong, J., Guild, F.J., Ogin, S.L., Smith, P.A.: On matrix crack growth in quasi-isotropic laminates-I. Experimental investigation. Compos. Sci. Technol. **57**, 1527–1535 (1997)
2. Kenneth, L.R., Scott, W.C.: Damage Tolerance and Durability of Material Systems. Wiley, New York (2002)
3. Jean-Luc, R., Denys, G.: Strain energy release rate analyse of matrix micro cracking in composite cross-ply. Mater. Sci. Appl. **2**, 537–545 (2011)
4. McCartney, L.N.: Energy methods for fatigue damage modeling of laminates. Compos. Sci. Technol. **68**, 2601–2615 (2008)
5. Miami, P., Camanho, P.P., Mayugo, J.A., Turon, A.: Matrix cracking and delamination in laminated composites. Part II: evolution of crack density and delamination. Mech. Mater. **43**, 194–211 (2011)
6. Maria, K., Costas, S.: Stiffness and fracture analysis of laminated composites with off-axis ply matrix cracking. Compos. A: Appl. Sci. Manuf. **38**, 1262–1269 (2007)
7. Addis, T., William, M., Behrad, K., Suraj, R., Addis, K., Michel, V.T.: On the mechanical response of polymer fiber composites reinforced with nanoparticles. Mech. Compos. Multi-Funct. Mater. **7**, 125–130 (2016)
8. Addis, T., William, M., Behrad, K., Suraj, R., Addis, K., Michel, V.T.: Effects of nanoparticles on the shear properties of polymer composites. In: American Society of Composites-30th Technical Conference, 2015
9. Suraj, R., Addis, T., Addis, K., Michael, A.S.: Meso-scale deformation mechanisms of polymer bonded energetic materials under dynamic loading. Mech. Compos. Multi-Funct. Mater. **7**, 451–456 (2016)
10. Suraj, R., Addis, T., Addis, K.: Local deformation and failure mechanisms of polymer bonded energetic materials subjected to high strain rate loading. J. Dyn. Behav. Mater. (2016). doi:10.1007/s40870-016-0051-9
11. Behrad, K., Suraj, R., Addis, K.: Meso-scale strain localization and failure response of an orthotropic woven glass–fiber reinforced composite. Composites Part B **78**, 308–318 (2015)
12. Canal, L.P., Gonzalez, C., Molina-Aldareguia, J.M., Segurado, J., LLorca, J.: Application of digital image correlation at the microscale in fiber-reinforced composites. Compos. A: Appl. Sci. Manuf. **43**, 1630–1638 (2012)
13. Jay, D.C., Wael, A., John, L., Huseyin, S.: High resolution digital image correlation measurements of strain accumulation in fatigue crack growth. Int. J. Fatigue **57**, 140–150 (2013)

# Chapter 6
# Experimental Investigation of Strength of Curved Beam by Thin Ply Non-Crimp Fabric Laminates

**M.A. Arca, M. Papila, and D. Coker**

**Abstract** Resistance against delamination failure and through the thickness tensile properties of curved carbon fiber reinforced plastics composites are investigated experimentally by conducting the curved beam strength tests. Effect of novel material thin ply non crimp fabric (NCF) architecture on delamination resistance of carbon fiber reinforced composites are investigated and compare with that of standard UD layups. In order to determine through the thickness tensile properties of curved carbon fiber composites, standard test method is carried out, namely four-point bending tests. The dynamic delamination propagation and failure sequences under curved beam bending is captured using Photron© Fastcam SA5 ultra high speed system. For the non-crimp fabric configuration an increase in the curved beam strength is observed in comparison with [0] and [0/45/-45/0] laminates by unidirectional (UD) tape material. For the UD tape, the initial defects caused by the out-of-autoclave manufacturing process is found to be the potential failure sites. The test results and observations suggest that thin-ply NCF is much less vulnerable to the existence of manufacturing voids in contrast to standard thickness UD tape. Finally, TPNCF is shown to have superior properties in regard to delamination resistance and curved-beam strength.

**Keywords** Delamination • Curved composite • Thin-ply non-crimp fabric • High-speed camera • Curved beam strength

## 6.1 Introduction

Composite materials have superior properties compared to metals such as improved strength, stiffness, fatigue and impact resistance, thermal conductivity, corrosion resistance [1]. Light weight structures with improved properties of composite materials are required in advanced technology industries such as aerospace and wind energy, leading to increase in the demand for these materials especially carbon fiber reinforced plastics (CFRP). On the other hand, properties of fiber reinforced laminated composites depend highly on manufacturing methods and material quality [2–4]. The significance of these factors is much higher in case of complex shaped composite laminates. Load carrying capacity at the radius of cornered structures for instance may be susceptible due to risk of delaminations, the separation of layers with significant loss of mechanical integrity and toughness [1].

Development of the novel spread tow-thin ply technique eventuated in new class of material called thin ply non-crimp fabric (TPNCF). Thin ply non crimp fabric is a class of material with multiple layers each as half the thickness of conventional UD tape. Its structure with no crimp provides more homogeneity and closer performance to UD materials. The concept of non-crimp fabric or multiaxial composites has been utilized since 1990s by basically stitching the conventional unidirectional reinforcement layers in different orientations to create fabric like material with minimal fiber waviness. Pre-assmbly of possible orientations may reduce the cost and time required for manufacturing of the composite laminates compared to use of unidirectional plies indivudualy. Albeit the reduced in plane properties of the non-crimp fabric composites compared to equivalent UD configuration, they provide reduced waviness and increased mechanical properties in the out-of-plane and impact performance over traditional fabric type composites. Besides, stitching can improve through

M.A. Arca
Department of Aerospace Engineering, Middle East Technical University, 06800 Ankara, Turkey
e-mail: mirayarca@gmail.com

M. Papila
Materials Science and NanoEngineering, Sabanci University, 34956 Istanbul, Turkey

D. Coker (✉)
Department of Aerospace Engineering, Middle East Technical University, 06800 Ankara, Turkey

METU Center for Wind Energy, Middle East Technical University, 06800 Ankara, Turkey
e-mail: coker@metu.edu.tr



A.T. Zehnder et al. (eds.), *Fracture, Fatigue, Failure and Damage Evolution, Volume 8*,
Conference Proceedings of the Society for Experimental Mechanics Series, DOI 10.1007/978-3-319-42195-7_6

the thickness strength, delamination and impact resistance [5, 6]. Since early 2000s development of the novel spread tow-thin ply technique has been adopted into fiber reinforcement alternatives. This method enables producing dry ply thicknesses as low as 0.02 mm. Thin plied NCF forms are also made available by stitching the tow spread thin UD layers in the desired configuration instead of weaving the filaments [7, 8].

In this study, Failure behavior of thin ply non-crimp fabric laminates are determined by conducting ASTM standard fracture toughness and curved beam strength tests. The results are compared with the equivalent composite laminates of the lay-up made by stacking conventional UD plies. High speed camera system is used in the experiments in order to gain better understanding of the effect of thin ply non-crimp fabric materials on failure process and the dynamic crack growth process. Micrographs of microstructures are examined and compared.

## 6.2 Material

In this study composite laminates are made of prepreg of Choromat C-Ply (made of Toray T700 carbon fiber tows) TPNCF and UD tape prepreg of standard ply thickness (nominally 0.125 mm). The resin is ALDILA AR 2527 toughened epoxy. The C-Ply NCFs have 0.0625 mm ply thickness which is half of the UD tape prepreg ply thickness.

Thin ply NCF laminates are composed of 24 layers and each NCF layer is formed by two thin plies tied together. The tied/stitched prepreg NCF is in the [0/45] configuration. TPNCF laminates were made as $[0/45/\text{-}45/0]_{12T}$ which is equivalent to stacking of 48 thin plies. Equivalent laminates of standard UD tape are composed of 24 plies, stacked as $(0/45/\text{-}45/0)_{6T}$. Figure 6.1 represents the microscopic cross-section views of the thin-ply NCF and UD tapel made laminates.

L-shaped beams and flat plates were manufactured for test specimen preparation. Firstly tool was covered with the release film then prepreg plies were laid up on the tool followed by a layer of peel ply and breather. The layup on the tool was then vacuum bagged and cured at 135 °C for 6 h. After the curing process L-shaped beams and laminates were cut into specimens by Diatrim diamond saw. Specimen geometries and dimensions are shown in Fig. 6.2a, b, respectively. Thin-ply non-crimp fabric and UD tape made specimens were about 3.5 mm thick. L-shaped specimen's leg lengths were 100 mm with inner radius of 10 mm.

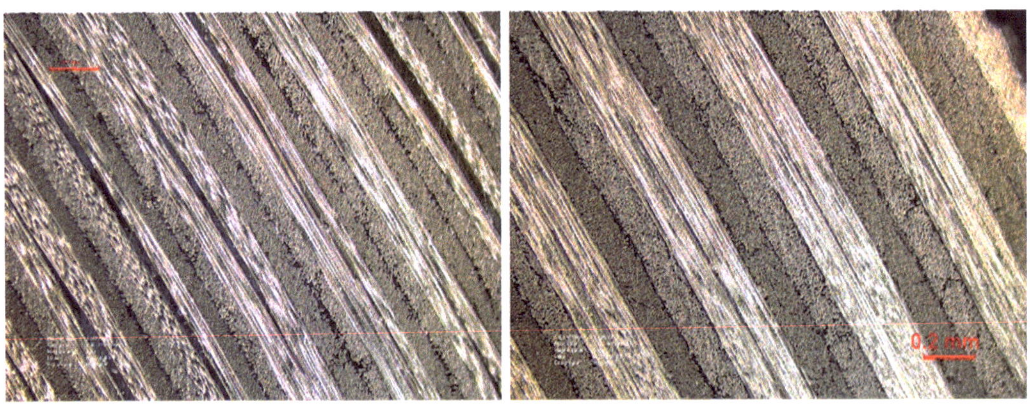

**Fig. 6.1** Microscopic side view of [0/45/-45/0] sequenced laminates by: (**a**) thin ply NCF, (**b**) conventional UD tape

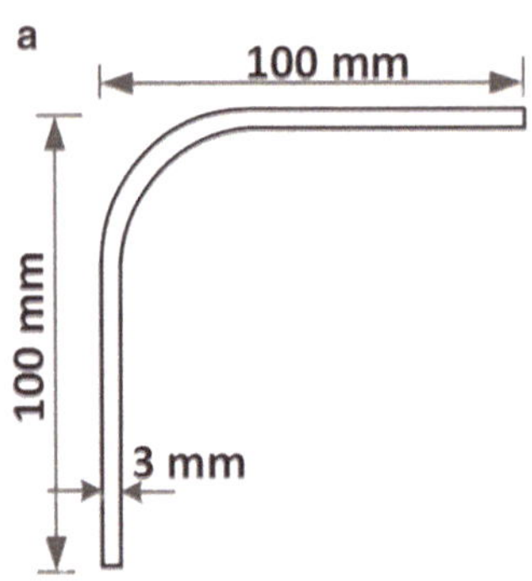


b

| | TPNCF | UD |
|---|---|---|
| L1=L2 | 100 | 100 |
| w (mm) | 25 | 25 |
| t (mm) | 3.5 | 3.5 |
| Ply # | 24 | 12 |
| # of specimens | 3 | 3 |

**Fig. 6.2** (**a**) Typical specimen geometry. (**b**) tested specimen dimensions

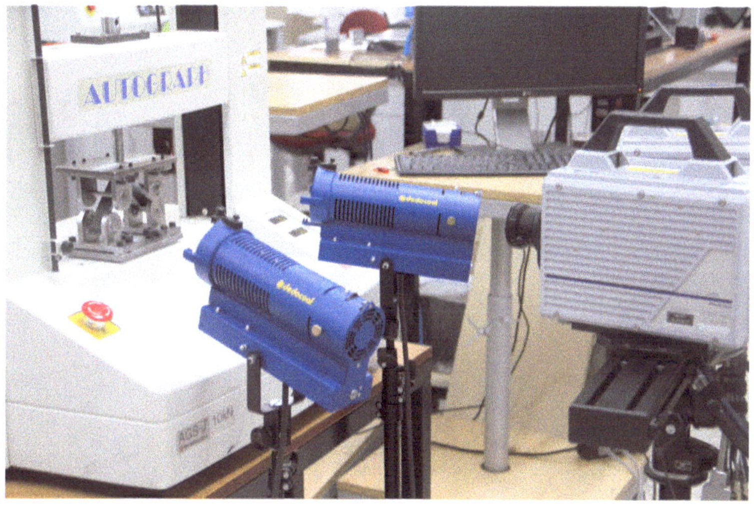


**Fig. 6.3** Four-point bending test—setup schematic showing test fixture, lighting and high speed camera

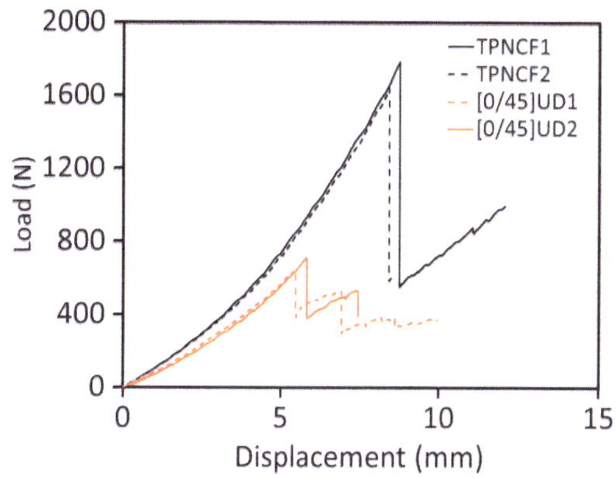


**Fig. 6.4** Four-point bending load displacement curves of [0/45/-45/0] sequenced UD tape and thin ply NCF made specimens

## 6.3 Experimental Method

ASTM D6415 standard four-point bending tests aim to determine curved beam strength of 90° curved composite specimens [9]. The tests were conducted using a 10 kN Shimadzu Autograph AGS-J screw driven tensile testing machine and the load displacement data was recorded. The curved beam strength of a composite laminate is determined by the moment per unit width which generates the delamination failure. In the test, constant bending moment is applied to the specimen corner through the arms. Experimental setup configuration used in the four-point bending tests is shown in Fig. 6.3. During the test Photron ASA5 high speed camera (HSC) is used for recording delamination propagation. Edge of the specimen is painted with white paint to make failure (crack) visible. Specimen is loaded at 0.5 mm/min crosshead speed according to the ASTM D6415 test standard. End trigger mode of HSC is used and photos are captured by manually triggering the camera when the cracking sound is heard which corresponds to the load drop in the load displacement curve. Test is stopped after 50 % load drop of the max load occurs according to the standard.

## 6.4 Results

Load displacement curves of two TPNCF and two UD tape made specimens of repeated [0/45/-45/0] sequence recorded from the four-point bending tests are presented in Fig. 6.4. UD tape made specimens resisted up to 700N and multiple load drops were observed. After the first load drop, slope did not change, only small load drops were observed, which indicates stable crack growth and partial loss of load carrying capacity. After the second load drop, specimens lost the resistance to crack growth. Thin ply non-crimp fabric made specimens' load displacement behavior was different from the equivalent UD material case. The stiffness and load carrying capacity are higher TPNCF made specimens carried load as high as 1800N and

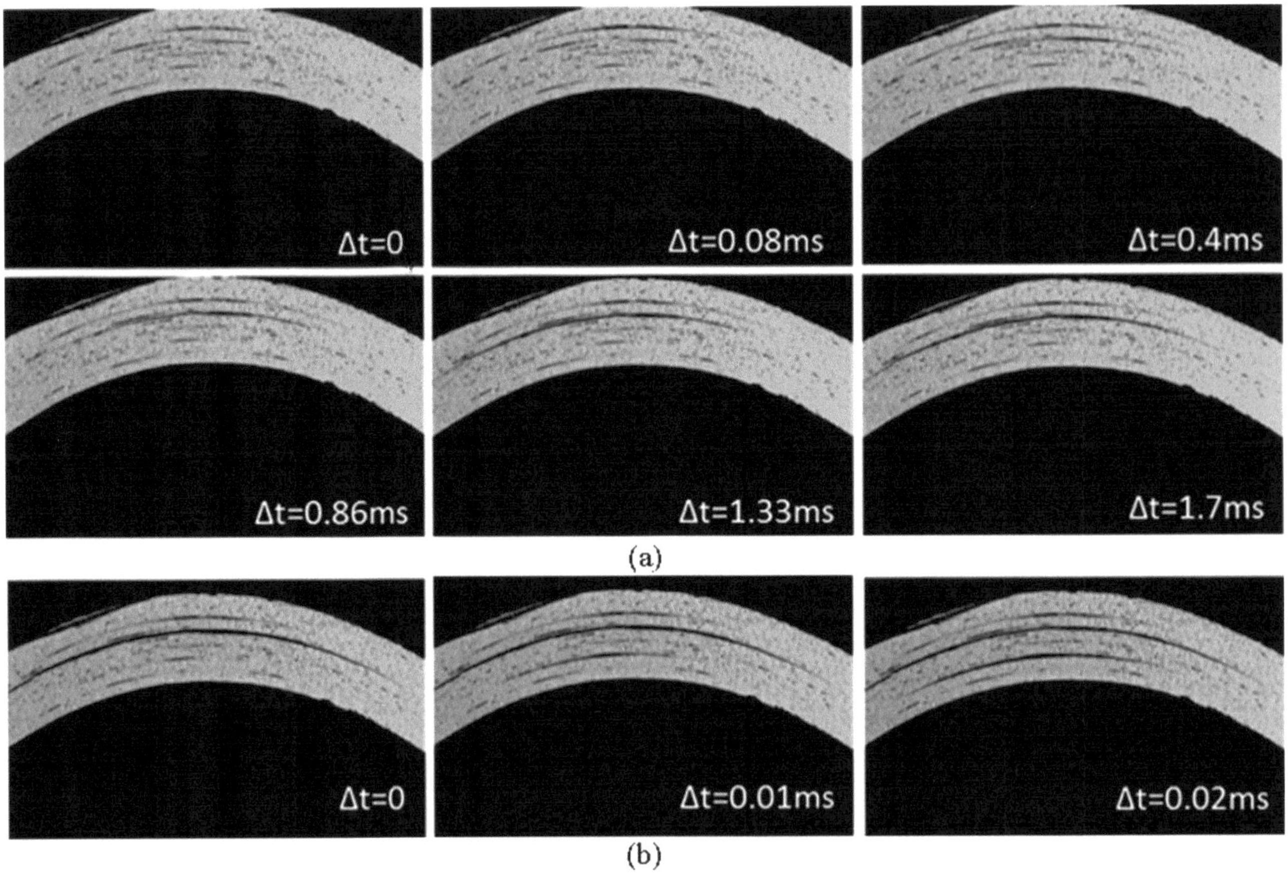


**Fig. 6.5** High speed camera images taken at 100,000 fps for specimen UD2 showing delaminations associated with: (**a**) the first load drop, (**b**) the second load drop

one load drop was observed. Load drop was down to 500N, indicating that the failure occured throughout the width direction and specimens lost the stiffness and load carrying capacity.

Snapshot photos from the high speed camera of UD2 laminate at first and second load drops taken at 100,000 fps are given in Fig. 6.5a, b, respectively. First picture of Fig. 6.5a ($\Delta t = 0$) shows the specimen at 370N prior to the first load drop. Initial defects can be seen clearly at the curved section throughout the thickness. After 80 μs (second picture) first delamination started from the initial defect close to the outer radius. Up to 0.4 ms the first delamination propagated slightly and at 0.4 ms second delamination initiated from another existing defect close to the mid-plane. At the first load drop whole process ended in 1.7 ms and the second delamination growth was very stable. Next load drop is when the second delamination propagated to the arms along with a a third delamination nucleated which also propagated slightly from a small defect, but close to inner radius. Whole process ended in 0.02 ms. For both of the load drops failure process was considerably long in ms level, indicating low cracking speeds. Between the first and second load drop smaller load drops were also observed in the load displacement curves, however no crack initiation and/or propagation are evident when the last picture of the first load drop in Fig. 6.5a and the first picture of second load drop in Fig. 6.5b are compared. This suggests that the cracks do not extend throughout the width of the specimen and during these small load drops crack propagation occurs through the width direction. This argument is supported by comparing the front and back sides of the specimen where different failure patterns are observed for both UD laminates.

Figure 6.6 shows the failure sequence of thin fiber NCF specimens in one load drop captured by high speed cameras at 100,000 fps. The first picture of the Fig. 6.6 ($\Delta t = 0$) represents the initial state of the specimen prior to failure at 1700N. The last picture in Fig. 6.6 shows the specimen at the end of the process which about takes 70 μs in total. In the second picture it is seen that a single major delamination at the mid-plane initiated and propagated through the arms in 10 μs. In the same picture slight propagation of a second delamination two ply below the first delamination and a third delamination between middle and inner radius at the right half of the curved region in first 10 μs was observed. A fourth relatively small delamination occured between the middle and outer radius from the left side above the mid-plane of the curved region

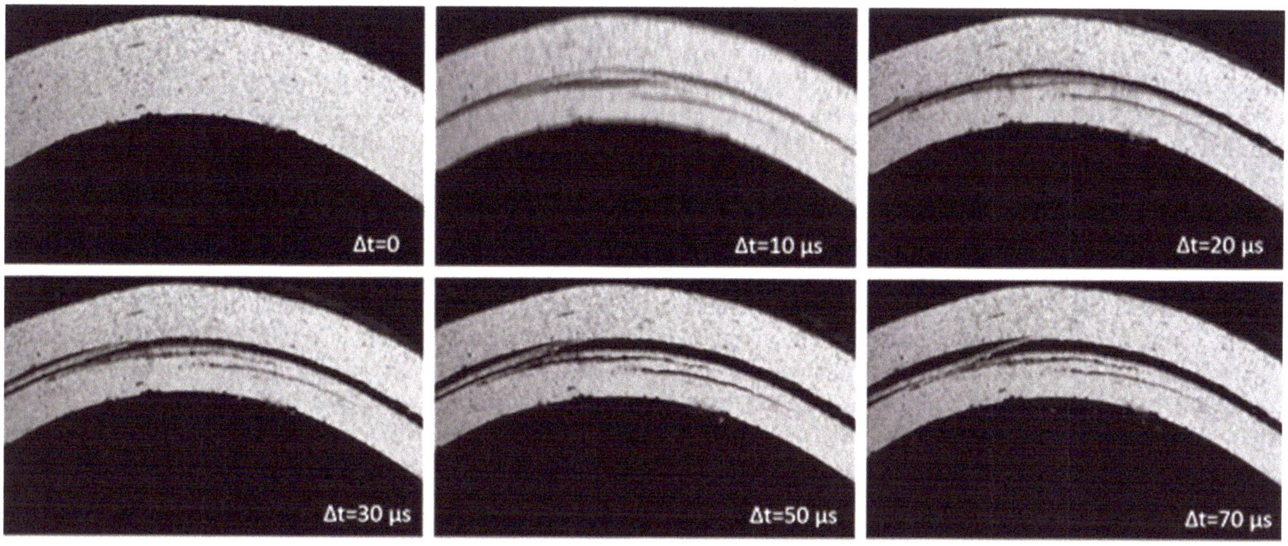


**Fig. 6.6** High speed camera photos showing failure sequence for TPNCF specimen at 100,000 fps

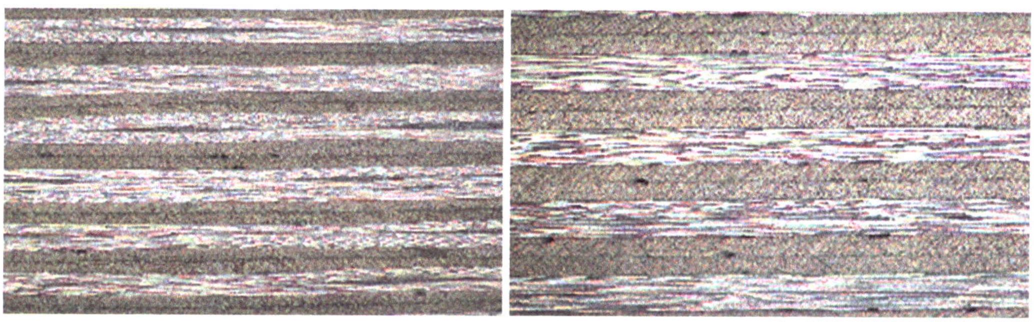

**Fig. 6.7** Sideview Micrographs of (**a**) TPNCF made specimen, (**b**) UD tape made specimen showing the small initial defects close to the curve region

(between 10 and 20 μs). For the next 50 μs, first delamination propagated to the curved part. For all of the tests two major delaminations were observed at the mid part of the specimen and failure process ended in nearly 70 μs. Besides, for all tests the first delamination initiated and propagated through the arms in 10 μs dynamically.

Figure 6.7 below shows the micrographs of TPNCF and UD tape made specimens at the arm region very close to the curved regions, respectively. When Fig. 6.7 is compared with Fig. 6.2, small voids are observed for both type of materials, however at the curved regions larger initial defects are evident which are the root cause for the initiation of small delaminations shown in the HSC photos. The high speed photos taken for all UD specimens show that initial defects at the curved region are the failure initiation sides. However, even though small voids are also observed for TPCNF laminates as mentioned above, their presence is not fatal for the specimen, and neither load displacement curve nor the failure pattern is affected. These test results and observations suggest that thin-ply NCF is much less vulnerable to the existence of manufacturing voids in contrast to standard thickness UD tape. Thus, TPNCF is shown to have superior properties in regard to delamination resistance and curved-beam strength.

## 6.5 Conclusions

In this study, curved beam strength and failure process of thin-ply non-crimp fabric material are investigated and compared with that of the standard thickness UD tape based laminates. The laminates were manufactured by hand layup and vacuum bagging of the prepreg forms of these materials. ASTM standard four-point bending tests were conducted and the dynamic delamination propagation and failure sequences under bending loading were captured by high speed camera system for comparative investigation. UD tape made specimens were found to resist up to 700N with two load drops observed, where cracks nucleated from the initial defects. The argument that cracks do not extend all through the width of the specimen and specimen does not lose its load carrying capacity completely after the load drop is attributed to the different failure surfaces at the front and back sides of a single specimen. Moreover different failure patterns on different UD made specimens were observed by analyzing the high speed camera photos taken during the failure process. For the thin ply non-crimp fabric laminates it was observed that one load drop occured and after the load drop the specimen loses its stiffness and load carrying capacity. High speed camera photos show that two major delaminations occur at the mid-plane of the specimen. When failure process of TPNCF and UD laminates are compared, failure process is very long for UD laminate and crack lengths are small compared to TPNCF laminate. For the TPNCF laminate first delamination initiated and propagated through the arms in 10 μs, which was dynamic. Different failure load, location and sequence for UD and TPCNF laminates are attributed to stitching and the thickness of the TPNCF laminates since manufacturing and test conditions are the same. In conclusion, it is thought that the thin-ply non-crimp fabric type material increases the curved beam strength since stitching provides extra strength and also thin fiber provides more homogeneity resulting in enhanced properties. Furthermore, thin ply non-crimp fabric type material is found to be easy to process than the UD laminates under the same manufacturing steps, and more suitable for production of complex geometries.

**Acknowledgements** This work was supported by the Ministry of Science, Industry and Technology and Turkish Aerospace Industries through grant 00785.STZ.2011-1.

## References

1. Kaw, K.A.: Mechanics of Composite Materials. CRC Press/Taylor & Francis Group, Boca Raton/London/New York (2006)
2. Sørensen, B.F., Jørgensen, E., Debel, C.P., Jensen, F.M., Jensen, H.M., Jacobsen, T.K., Halling, K.M.: Improved design of large wind turbine blade of fibre composites based on studies of scale effects (Phase 1)—Summary Report, Risø National Laboratory, Denmark, Risø-R-1390 (EN) (2004)
3. Edwards, T., Thompson, J.: Spar corner radius integrity for the A400M wing. Appl. Mech. Mater. **3–4**, 197–204 (2005)
4. Vanttinen, A.: Strength prediction of composite rib foot corner. Ms Thesis, Helsinki University of Technology (2008)
5. Arteiro, A.J.C.: Technology development and structural mechanics of composites built of spread tow thin-ply technology. Ms Thesis, University of Porto (UP) (2012)
6. Roure, T., Sanial, P.: C-PLY™, a new structural approach to multiaxials in composites. JEC Compos. Mag. (68), 53–54 (October 2011). Special JEC Asia
7. Shin, S., Kim, R.Y., Kawabe, K., Tsai, S.W.: Experimental studies of thin-ply laminated composites. Compos. Sci. Technol. **67**, 996–1008 (2007)
8. Tsai, S.W., Nettles, A.T.: Representative test data on bi-angle thin-ply NCF. JEC Compos. Mag. **68**, 62–63 (2011)
9. ASTM International: Standard Test Method for Measuring the Curved Beam Strength of a Fiber-Reinforced Polymer-Matrix Composite. ASTM D6415/D6415M – 06a. ASTM International, West Conshohocken, PA (2007)

# Chapter 7
# Role of Laminate Thickness on Sequential Dynamic Delamination of Curved [90/0] CFRP Composite Laminates

**Imren Uyar, B. Tasdemir, D. Yavas, and D. Coker**

**Abstract** In aerospace industry, high demand for the lightweight structures are fostering the use of carbon fiber reinforced polymer composites in a wide variety of shapes, as primary load carrying elements. However, once a composite laminate takes a highly curved shape, such as an L-shape, interlaminar stresses augmented in the curved region cause highly dynamic delamination nucleation and propagation. This paper provides experimental observations of dynamic delamination failure in cross-plied L-shaped composite laminates under quasi-static shear loading for varying laminate thickness. In the experiments, load-displacement curves are recorded and dynamic delamination events areas captured using a million fps high speed camera. In our previous work, two distinct types of failure modes have been identified depending on the laminate layup: (i) formation of multiple delaminations leading two single load drop in its load-displacement curve during the failure of unidirectional laminates, $[0]_{17}$, and (ii) formation of sequential delaminations associated with each discrete load drop in its load-displacement curve were during the failure of cross-ply laminates, $[90/0]_{17}$. Accordingly this current study shows that formation of sequential delaminations is independent from the laminate thickness.

**Keywords** Delamination • L-shaped laminates • Dynamic crack growth • High-speed camera

## 7.1 Introduction

Carbon fiber reinforced polymer (CFRP) composites have been extensively utilized in both aerospace and wind turbine industries because of their high strength–to-weight ratio, over the last few decades. Many structurally critical metallic airframe components have been replaced by their composite substitutes, such as L-shaped joints within a typical wing box. Such replacement showed that the material response and fracture modes are highly different for the composite components due to their anisotropic nature, though they are subjected to the similar external loading. The loading in a typical L-shaped joint in a wing box can be reduced to three simple loading cases: (i) axial load which is parallel to the arm (P), (ii) shear load which is perpendicular to the arm (V), and (iii) the moment (M) as shown in (Fig. 7.1a). These external loads result in delamination in the composite laminates in the curved region due to the low through-the-thickness strengths of laminated composites. The delamination failure of L-shaped composite laminates under perpendicular loading to the arm was firstly examined by Martin et al. [1, 2], wherein the numerical study showed that delamination initiates at the location of highest radial stress in curved region. They also showed that delamination propagates in to the arms of the laminate predominantly in opening mode with the help of energy release rate analysis, though the unstable the growth of delamination was not captured Thereafter, Wimmer et al. [3] numerically showed the stages of the unstable delamination propagation by finite element method employing virtual crack closure technique. In addition, in their experimental work, they found an instantaneous load drop in load displacement curve accompanied by a delamination propagation. Then Feih and Shercliff (2005) [4] analyzed the failure of L-shaped composite laminates positioned between a composite base and vertical rib for perpendicular to the arm loading case. The finite element model was performed on ABAQUS with UMAT subroutines combined with Hashin's failure criteria for matrix and fiber cracking and with Tong-Norris delamination onset criterion. In numerical results, they indicated the failure sequence and failure types with locations and validated by strain gage results in experiments. Gozluklu

I. Uyar
Department of Aerospace Engineering, Middle East Technical University, 06800 Ankara, Turkey

B. Tasdemir • D. Coker (✉)
Department of Aerospace Engineering, Middle East Technical University, 06800 Ankara, Turkey

METU Center for Wind Energy, Middle East Technical University, 06800 Ankara, Turkey
e-mail: coker@metu.edu.tr

D. Yavas
Department of Aerospace Engineering, Iowa State University, Ames, IA 50011, USA



A.T. Zehnder et al. (eds.), *Fracture, Fatigue, Failure and Damage Evolution, Volume 8*,
Conference Proceedings of the Society for Experimental Mechanics Series, DOI 10.1007/978-3-319-42195-7_7

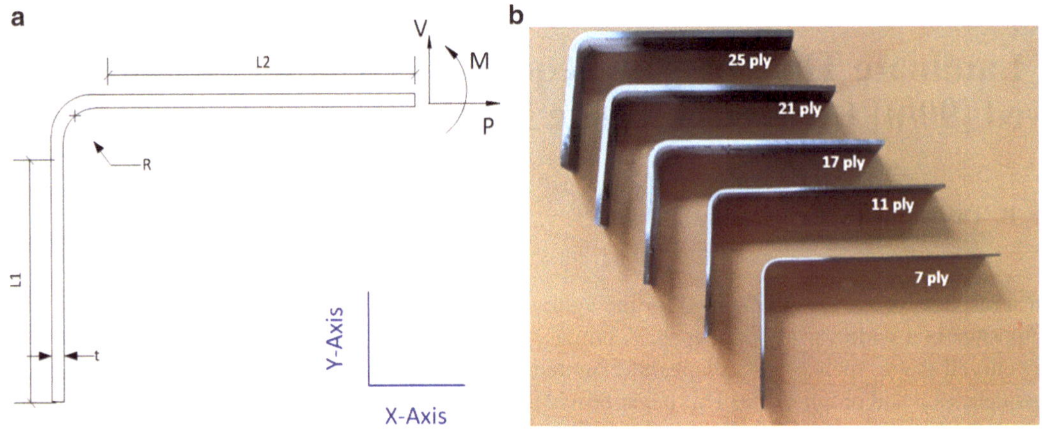


**Fig. 7.1** (**a**) Loads applied on L-beam structure and the symbols of dimensions. (**b**) Specimens with variable thicknesses

**Table 7.1** Orientation, ply numbers, and geometrical constraints of specimens using in shear loading experiments

| Orientation | # of Plies | $t$ (mm) | $\boldsymbol{R_i}$ (mm) | L1 (mm) | L2 (mm) | $w$ (mm) |
|---|---|---|---|---|---|---|
| $\left[(90/0)_4, \overline{90}\right]_s$ | 17 | 3.18 | 10 | 90 | 150 | 30 |
| $\left[(90/0)_5, \overline{90}\right]_s$ | 21 | 3.95 | 10 | 90 | 150 | 30 |
| $\left[(90/0)_6, \overline{90}\right]_s$ | 25 | 4.67 | 10 | 90 | 150 | 30 |
| $[90/0]_{11}$ | 11 | 2.10 | 10 | 90 | 150 | 30 |
| $[90/0]_7$ | 7 | 1.33 | 10 | 90 | 150 | 30 |

and Coker (2012) [5] carried out explicit finite element analysis with cohesive elements to model delamination in composite L-beams subjected to parallel loading instead of perpendicular loading. In their simulations, they observed dynamic crack growth and the crack tip speed reaches the shear wave speed of the laminate.

In our previous work [8], we have showed that there are two distinct mechanisms leading to delamination failure of $[0]_{17}$ and $[90/0]_{17}$ lay-up L-shaped composite laminates. The first is the formation of multiple delaminations in $[0]_{17}$ lay-up, accompanied by a sudden and large load drop in its load-displacement curve. The first delamination initiates on the mid-plane of the laminates, which is dynamically followed by two cracks nucleated symmetrically around each fronts of the initial delamination. Based on the analytical solution proposed by Lekhnitski [REF], the maximum radial stress under shear load occurs very near to the mid-plane of the laminates. Accordingly, the first delamination is initiated when the maximum radial stress reaches the interfacial strength of the laminates (~80MPa, [REF]) at the failure load. The second is the formation of sequential delaminations in $[90/0]_{17}$ layup, associated with multiple discrete load drops in its load-displacement curve. In the very early stage of the each load drop, the maximum circumferential stress forms matric cracks within the very inner 90° layer when it reaches the tensile strength of the lamina normal to the fiber direction (~81MPa, [REF]). Then these matrix cracks propagate towards the nearest interface between 0° and 90° layers and initiates the delamination propagating along this interface. Such mechanism is consecutively repeated for the each load drop observed in the load-displacement curve. In the current study, we explore the role of the laminate thickness on the sequential delamination mechanism observed for cross-ply L-shaped composites.

The effective on the repetitiveness of the failure mechanism is considered the thickness. Therefore, the effect of geometrical constraints are observed in this study. For the test specimens, different ply numbers are given which are 7, 11, 21 and 25 plies (Fig. 7.1b). The results are compared with the 17-ply 90/0 laminate.

## 7.2 Experimental Method

L-shaped samples were constructed by staking HexPly® AS4/8552 UD carbon pre-pregs using right angled male tool. The number of plies varies depending on the desired laminate thickness as shown in Table 7.1. Standard fabrication of vacuum bagging and autoclave curing method were implemented. Figures 7.1a, b shows the samples geometry, and the

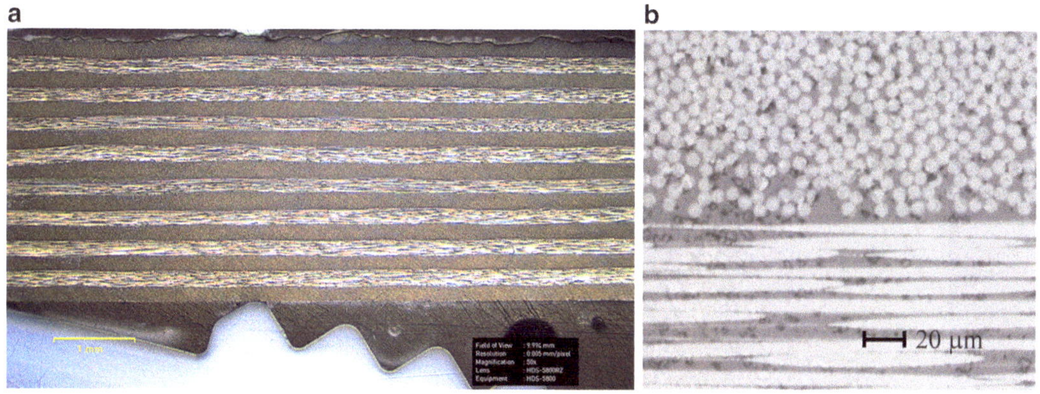


**Fig. 7.2** (**a**) An optical micrograph of composite specimens showing 90° and 0° degree plies for $[90/0]_{17}$ layup, (**b**) A repeat unit cell showing fibers parallel (90°) and normal (0°) to the plane

**Table 7.2** Lamina properties [7]

| | $E_1$ (GPa) | $E_2 = E_3$ (GPa) | $G_{12} = G_{13}$ (GPa) | $\nu$ | Ply thickness (mm) | Density (gr/cm$^3$) | Nominal fiber volume (%) |
|---|---|---|---|---|---|---|---|
| AS4/8552 UD | 135.0 | 8.5 | 4.2 | 0.29 | 0.184 | 1.59 | 57.42 |

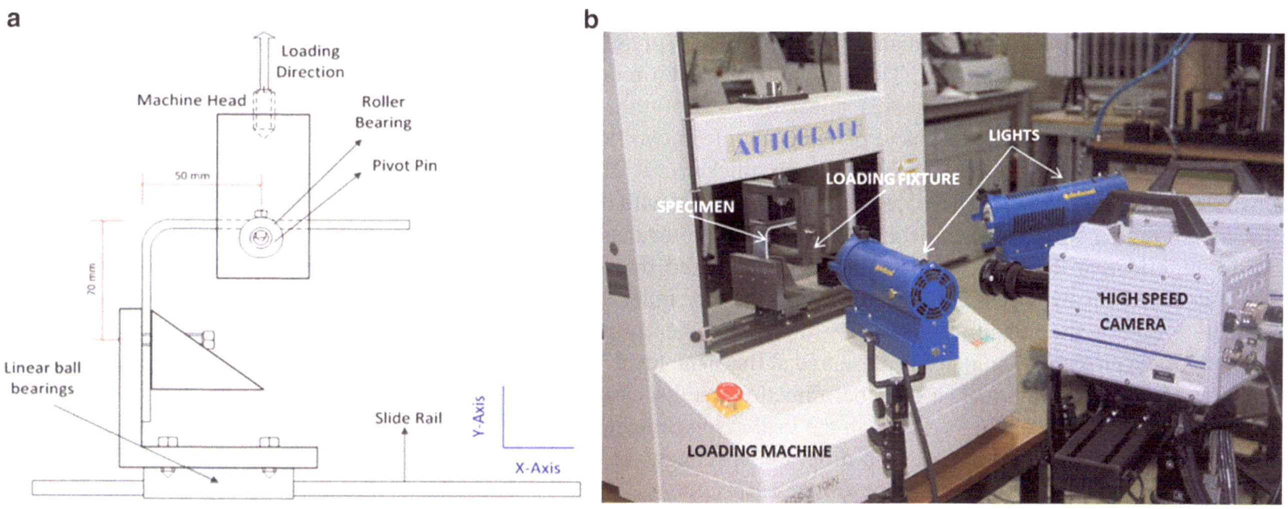


**Fig. 7.3** (**a**) Loading fixture for perpendicular loading of the arm; (**b**) Experimental setup showing Photron SA5 high speed camera and specimen after failure

corresponding dimensions are given in Table 7.1. In Figs. 7.2a, b optical micrographs are showing cross sectional distribution of carbon fibers in the epoxy matrix. Fibers are seen as white and matrix is grey (Table 7.2).

Shimadzu Autograph AGS-J series 10 kN screw-driven displacement controlled tensile-compression testing machine was used in the experiments. All the tests were conducted at a cross-head speed of 3 mm/min at the room temperature. The fixture was designed to introduce a displacement perpendicular to the upper arm. The direction of the applied displacement was maintained parallel to the lower arm during the experiment by means of a pivot pin-bearing mechanism attached to the upper arm of the sample. In addition, the samples were fastened to the pivot pin to fix the loading arm with respect to the corner of the samples. The lower arm of the specimen was also fastened to the lower part of the fixture. The lower part of the fixture was mounted on a linear motion bearing system to give smooth precision motion along the x-axis that removes the reaction force along the x-axis at the upper arm. Figures 7.3a, b schematically show a drawing of the loading fixture and the experimental setup, respectively. Utilization of a million fps high speed camera (Photron SA5) facilitated the observation of the dynamic delamination nucleation and propagation during the experiments. The captured images were used to determine the delamination growth speed.

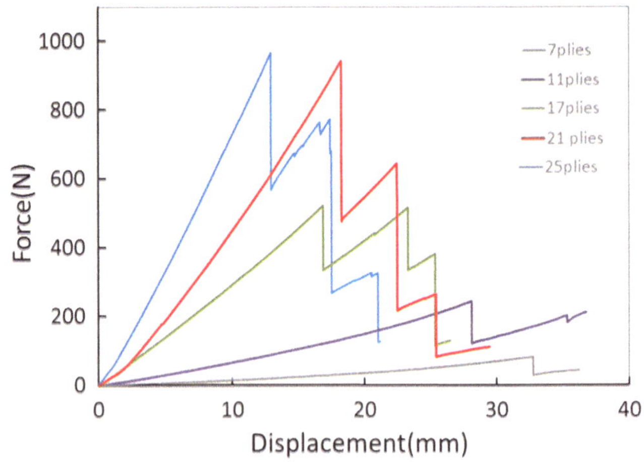


**Fig. 7.4** A representative set of force-displacement curves for varying laminate thickness

## 7.3 Results

Figure 7.4 depicts a representative set of force-displacement curves for the varying thickness. The increasing laminate thickness results in a gradual increase in the initial loading stiffness, as expected. Similarly, the failure load tends to increase for increasing laminate thickness, though the 21-ply and 25-ply exhibit the same level of failure load. However, it should be noted that the corresponding maximum displacement of 21-ply sample is around 30% higher than that of 25-ply sample due to its relatively higher compliance. Thinner samples exhibit much more compliant behavior. For example, the maximum load carried 7-ply specimen decreases down to 16% of the maximum load carried by 17-ply specimen, while the corresponding maximum displacement of 7-ply sample is two times bigger than that of 17-ply sample. One of the unique features observed during testing of the relatively thicker (17, 21, 25-ply) samples is the multiple load drops in the load-displacement curves. Each of these load drops is associated with the delaminations initiating sequentially in the inner part of the curved region. After of each delamination event, the residual load capacity of the samples is degraded by ~25–30 %, while the effective thickness of the samples is decreased by ~10–12 %. Based on the analytical solution proposed by Lekhnitski [REF], the circumferential stress in the very inner 90° layer of the samples remains almost constant with a ~25–30 % drop in the failure force and a ~10–12 % in the sample thickness. Accordingly, one can argue that the residual strength of the cross-ply laminates is preserved at the same level after each delamination event, which is roughly the tensile strength of the lamina normal to the fiber direction, as discussed previously. On the other hand, the thinner (7, 11-ply) samples exhibit only a single load drop in the load-displacement curves. The flexibility of the specimens is affected by the load carrying capacity. Up to first load drop, the specimens are very close to losing their curved shapes. After the first delamination growth, the stiffness values of these specimens drop to a lower value and lose their load carrying capacity.

Figure 7.5 show the normalized force and displacement curves. The force is normalized by square of the laminate thickness ($t^2$), which can be associated with bending stress, and the displacement is normalized by the tangent of the angle between the upper and lower arms of the sample prior to failure (Fig. 7.5a). Such normalization eliminates the scattering due to variation in the laminate thickness As can clearly be seen, the normalized loading stiffness is almost constant for varying laminate thickness. In addition, the normalized force is around 50 N/mm$^2$ within to be few percent for all the laminate thicknesses. However, delamination failure evolution after the first load drop exhibits different behavior for each case (Fig. 7.5b).

Figure 7.6 shows the delaminated samples after the first load drop in load-displacement curves for each laminate thickness. As can be seen, the first delamination nucleates in the very inner part of the curved region regardless of the laminate thickness. As discussed previously, very high circumferential stress in the inner of the laminate firstly nucleates matrix cracks within 90° layer, which is very weak along the circumferential direction. Thereafter these matrix cracks immediately propagate towards the interface between 0° and 90° layers, and initiate the interfacial delamination.

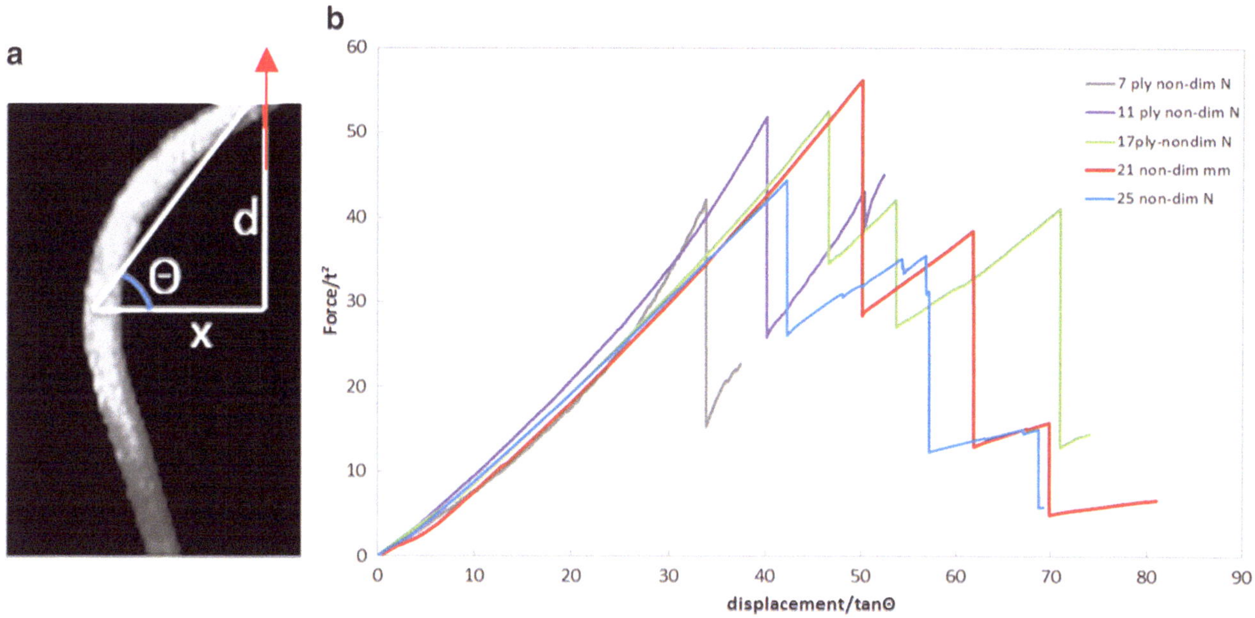


**Fig. 7.5** (**a**) The definition of the angle between the upper and lower arms of the loaded sample. (**b**) A replotting of Fig. 4 with the normalized load (by square of the laminate thickness) and displacement (by the angle between the upper and lower arms of the sample prior to failure) for varying laminate thickness

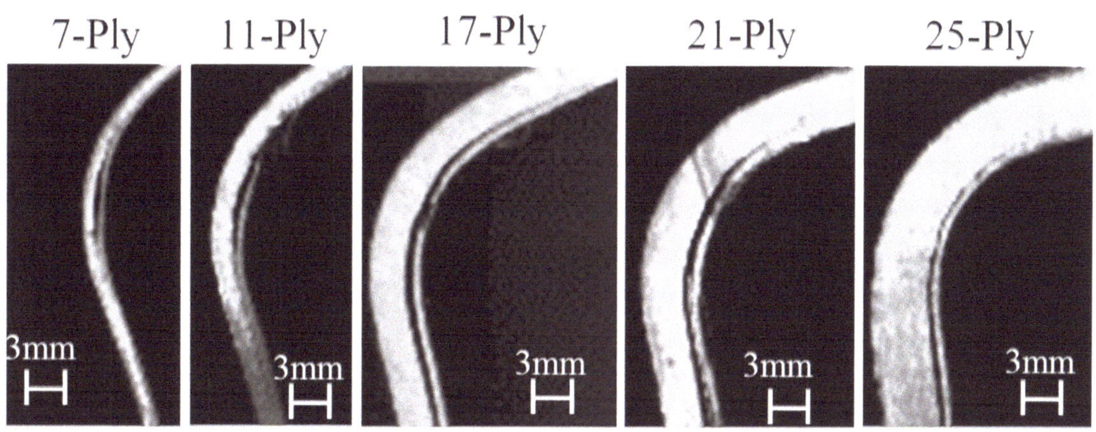


**Fig. 7.6** Comparison of initial delamination failure high speed pictures of [90/0] cross-ply specimens

## 7.4 Conclusions

In this study, dynamic delamination of L-shaped composite brackets with 90/0 layups with different ply numbers were investigated. For five different ply numbers, experiments were conducted in which L-shaped brackets were subjected to quasi-static shear loading. Subsequent dynamic delamination was captured with a high-speed camera and the load displacement curves were also recorded. High speed camera images proved that the failure mechanisms of specimen with different thicknesses are similar and the load displacement data shows that the normalized stiffness values are also the same. The major conclusions is that the failure mechanism of [90/0] cross-ply specimens under shear loading does not depend on the number of plies in the laminate.

## References


1. Martin, R.H., Jackson, W.C.: Damage prediction in cross-plied curved composite laminates. NASA Technical Memorandum 104089, USAAVSCOM Technical Report 91-B-009 (1991)
2. Martin, R.H.: Delamination failure in a unidirectional curved composite laminate. Compos. Mater. Test. **10**, 365–383 (1992)
3. Wimmer, G., Kitzmüller, W., Pinter, G., Wettermann, T., Pettermann, H.E.: Computational and experimental investigation of delamination in L-shaped laminated composite components. Eng. Fract. Mech. **76**, 2810–2820 (2009)
4. Feih, S., Shercliff, H.R.: Composite failure prediction of single-L joint structures under bending. Compos. Part A **36**, 381–395 (2005)
5. Gozluklu, B., Coker, D.: Modeling of the dynamic delamination of L-shaped unidirectional laminated composites. Compos. Struct. **94**, 1430–1442 (2012)
6. Uyar, I., Arca, M.A., Gozluklu, B., Coker D.: Experimental observations of dynamic delamination in curved [0] and [0/90] composite laminates. In: Carroll, J., Daly, S. (eds.) Fracture, Fatigue, Failure, and Damage Evolution, Volume 5: Proceedings of the 2014 Annual Conference on Experimental and Applied Mechanics, Conference Proceedings of the Society for Experimental Mechanics Series, pp. 189–196 (2015). doi:10.1007/978-3-319-06977-7__25
7. AS4/8552 Product Data. http://www.hexcel.com/Resources/DataSheets/Prepreg-Data-Sheets/8552_eu.pdf
8. Shivakumar, K.N., Allen, H.G., Avva, V.S.: Interlaminar tension strength of graphite/epoxy composite laminates. AIAA J. **32**(7), 1478–1484 (1994)

# Chapter 8
# Application of eMMC Model to Fracture of Metal Sheets

**Yueqian Jia, Sami Ghazali, and Yuanli Bai**

**Abstract** The fracture properties of TRIP780 and magnesium AZ31B-H24 alloy sheets were investigated in this paper. Mechanical experiments were performed for TRIP780 under uniaxial tension, notch tension, punch test, and plane strain tension. Mechanical experiments were performed for magnesium AZ31B-H24 sheets under different loading conditions, including monotonic uniaxial tension, notch tension, in-plane uniaxial compression, wide compression (or biaxial compression), plane strain compression, through-thickness compression, in-plane shear, punch test, and uniaxial compression-tension reverse loading. The stress invariants based Modified-Mohr-Coulumb (MMC) fracture model was transferred into an all-strain based MMC (eMMC) model under the plane stress condition, predicting the fracture strain in the space of strain ratio or $\Phi$ angle, instead of stress triaxiality and Lode angle parameter. The strain ratio or $\Phi$ angle could be directly measured by digital image correlation (DIC), while the latter required finite element analysis to be determined. This method made it possible to study fracture of materials while bypassing plasticity. Using the fracture strain measured by DIC, fracture locus was calibrated by the all-strain based MMC model. The fracture strain was extended by using a linear transformation operating to the plastic strain tensor to incorporate the anisotropic fracture behavior. Good prediction capability has been demonstrated for these two materials.

**Keywords** Modified-Mohr-Coulumb fracture model • All-strain based • Anisotropic fracture

## 8.1 Introduction

Sheet metal forming is an essential part of automobile industry because it allows for manufacturing high quality products with complex geometry with low costs. Recently, more and more lightweight materials like high strength steels or nonferrous metallic alloys (aluminum or magnesium) are adopted due to the urgent need to reduce vehicle weight and increase fuel efficiency and performance. However, these material are relative brittle and vulnerable to fracture during forming or subsequent services. It is necessary to have an adequate plasticity and fracture model to precisely predict the final shape without producing strains that exceeds the forming limit during metal forming. Many mechanical tests have revealed and proven that the material ductility is highly dependent on stress states, which are described by both stress triaxiality and the Lode angle parameter [1–3]. The Modified-Mohr-Coulumb (MMC) fracture model can successfully capture this dependency and predict the fracture locus using the stress parameters [4]. With the anisotropic effect in sheet metal fracture behavior, the relationship between stress state and fracture strain is usually difficult to be fully determined without a good plasticity model. Ground on this idea, the objective of this paper is to describe the anisotropic fracture properties by decoupling it from plasticity. Therefore, the MMC model is transformed into a space of strain states under the assumption of monotonic loading and Mises-Levy flow rule. The experimental results of two materials, TRIP780 steel sheets and magnesium AZ31B-H24 alloy sheets, are used to validate the strain-based MMC (eMMC) fracture criterion.

## 8.2 All Strain Based Anisotropic Ductile Fracture Modeling

The fracture locus of the original MMC fracture criterion reads

Y. Jia • S. Ghazali • Y. Bai (✉)
University of Central Florida, Orlando, FL, USA
e-mail: bai@ucf.edu



A.T. Zehnder et al. (eds.), *Fracture, Fatigue, Failure and Damage Evolution, Volume 8*,
Conference Proceedings of the Society for Experimental Mechanics Series, DOI 10.1007/978-3-319-42195-7_8

$$\overline{\varepsilon}_f(\eta,\overline{\theta}) = \left\{\frac{A}{C_2}\left[\tilde{C}_\theta^s + \frac{\sqrt{3}\left(C_\theta^{ax} - \tilde{C}_\theta^s\right)}{2-\sqrt{3}}\left(\sec\frac{\overline{\theta}\pi}{6} - 1\right)\right]\left[\sqrt{\frac{1+C_1^2}{3}}\cos\frac{\overline{\theta}\pi}{6} + C_1\left(\eta + \frac{1}{3}\sin\frac{\overline{\theta}\pi}{6}\right)\right]\right\}^{-\frac{1}{n}} \tag{8.1}$$

Where

$$C_\theta^{ax} = \begin{cases} 1 & \overline{\theta} \geq 0 \\ \tilde{C}_\theta^c & \overline{\theta} < 0 \end{cases}$$

And $\eta$ is the value of stress triaxiality, $\overline{\theta}$ is Lode angle parameter, $A$, $n$, $C_1$, $C_2$, $\tilde{C}_\theta^s$ and $\tilde{C}_\theta^c$ are the model parameters. The calibration is confined to the tensile loading condition because of the availability of the fracture strains. Due to the difficulty of measuring stress triaxiality evolution directly (without FE simulations with an adequate plasticity model), the model was transformed into an all-strain based space by involving the stress ratio and the ratio of in-plane principal strain increments [4, 5]. The strain incremental and stress ratio are defined as

$$\alpha = \frac{d\varepsilon_2}{d\varepsilon_1},\ \beta = \frac{\sigma_2}{\sigma_1} \tag{8.2}$$

Assuming the Mises-Levy flow rule, it can be derived that

$$\beta = \frac{2\alpha + 1}{2 + \alpha} \tag{8.3}$$

Note the relationship between $\eta$ and $\beta$ is

$$\eta = \frac{\beta + 1}{3\sqrt{\beta^2 - \beta + 1}} \tag{8.4}$$

Under plane stress condition, the Lode angle parameter $\overline{\theta}$ can be written as

$$\overline{\theta} = 1 - \frac{2}{\pi}arccos\left[-\frac{27}{2}\eta\left(\eta^2 - \frac{1}{3}\right)\right] \tag{8.5}$$

The fracture strain $\overline{\varepsilon}_f$ is hence expressed as a function of strain ratio ($\alpha$), which can be directly measured by a digital image correlating (DIC) system. However, the strain ratio ($\alpha$) could not distinguish different loading conditions between tensile dominated and compression dominated loadings. For example, it gives the same values for both equi-biaxial tension and equi-biaxial compression. In this manner, it is necessary to define another plastic strain based parameter to represent the effect of in-plane anisotropy under all the possible loading conditions. Therefore, the parameter $\Phi$ is used here and after, defined as

$$\Phi = atan2\left(-\dot{\varepsilon}_2^p, \dot{\varepsilon}_1^p\right) + 90^\circ \tag{8.6}$$

Where *atan*2 is the two arguments arctangent function with sign information included. The analogical equation in stress field can be used to calculate the Lode angle [6]. As a matter of fact, $\Phi$ denotes the angle to the positive minor strain direction in the forming fracture limit diagram (FFLD) under proportional loading, illustrated in Fig. 8.1. The anisotropic effect is thus measured quantitatively by the angle offset to the specified loadings under the assumption of isotropic J2 plasticity, which are 45° for equi-biaxial tension, 90° for plain strain tension, 116° for uniaxial tension, 135° for pure shear and 153° for uniaxial compression etc.

One big advantage of all-strain based fracture model is that it can be used to study fracture while bypassing plasticity. Recalling the definition of $\Phi$ under tensile loading condition

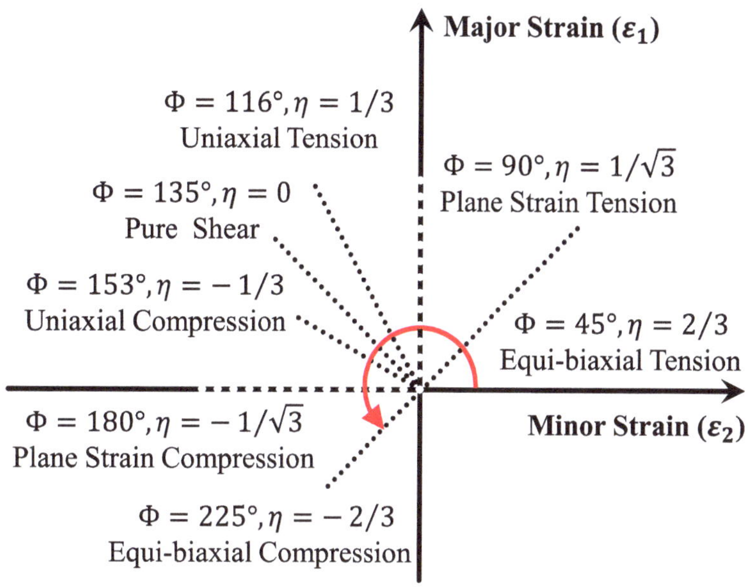


**Fig. 8.1** The schematic FFLD describes the definition of $\Phi$ under proportional loading. The symbol $\eta$ means the stress triaxiality

$$\Phi = atan2\left(-\dot{\varepsilon}_2^p, \dot{\varepsilon}_1^p\right) + 90^\circ = 2\,arctan\frac{-\dot{\varepsilon}_2^p}{\sqrt{\left(\dot{\varepsilon}_2^p\right)^2 + \left(\dot{\varepsilon}_1^p\right)^2} + \dot{\varepsilon}_1^p} + 90^\circ = 2\,arctan\frac{-\alpha}{\sqrt{1+\alpha^2}+1} + 90^\circ \tag{8.7}$$

The fracture strain $\overline{\varepsilon}_f$ is therefore expressed as a function of $\Phi$ only, by Eqs. (8.1)–(8.7), which becomes an all-strain based fracture locus. Furthermore, in order to describe the anisotropic fracture accurately, a non-conjugated anisotropic equivalent plastic strain function is included by operating on a linearly transformation to the strain tensor [7]. In the present work, the anisotropic equivalent plastic strain is defined as

$$\overline{\varepsilon}_p = \sqrt{\frac{2}{3}(\boldsymbol{\beta}\boldsymbol{\varepsilon}^p : \boldsymbol{\beta}\boldsymbol{\varepsilon}^p)} \tag{8.8}$$

Where $\overline{\varepsilon}_p$ indicates the plastic strain vector under plane stress condition $\boldsymbol{\varepsilon}^p = \left\{\varepsilon_{11}^p, \varepsilon_{22}^p, \varepsilon_{33}^p, \varepsilon_{12}^p\right\}$, while $\boldsymbol{\beta}$ is a positive semi-definite matrix which characterizes the linear transformation of the strain vector, reads

$$\beta = \begin{bmatrix} \beta_{11} & 0 & 0 & 0 \\ 0 & \beta_{22} & 0 & 0 \\ 0 & 0 & \beta_{33} & 0 \\ 0 & 0 & 0 & \beta_{12} \end{bmatrix} \tag{8.9}$$

Where $\beta_{11}$, $\beta_{22}$, $\beta_{33}$, and $\beta_{12}$ are four anisotropic fracture coefficients. Now the new fracture model reads

$$\overline{\varepsilon}_p \le \overline{\varepsilon}_f\left(\eta, \overline{\theta}\right) = \hat{\varepsilon}_f(\alpha) = \tilde{\varepsilon}_f(\Phi) \tag{8.10}$$

Therefore the experimental data were rotated back into the sheet rolling orientation, the coefficients of both fracture locus and the anisotropic equivalent strain were then calibrated.

## 8.3 Simulation Results

In this section, the eMMC model is implemented into a user-subroutine (VUMAT) in Abaqus/Explicit, with anisotropic yield criterion and hardening model. The experimental data from two example materials are used to validate the eMMC fracture locus: TRIP780 steel sheet (see Figs. 8.2 and 8.3; Table 8.1) and magnesium AZ31B-H24 alloy sheet (see Figs. 8.4 and 8.5; Table 8.2) under various loading conditions.

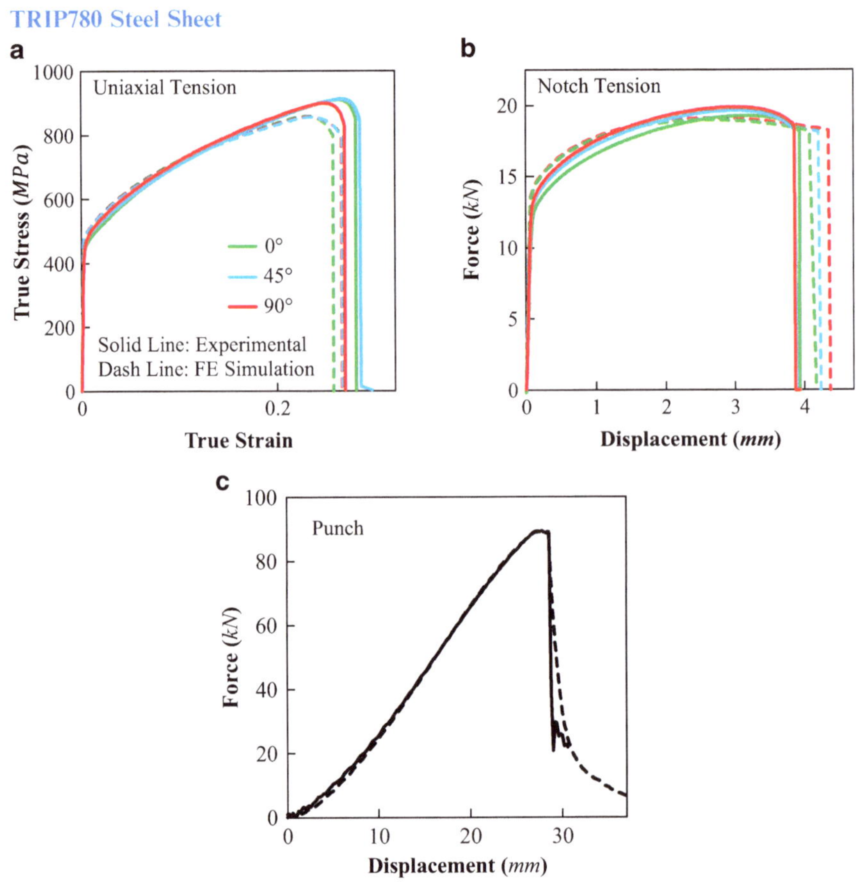


**Fig. 8.2** Comparison between tested and simulated results for TRIP780, in (**a**) true stress-strain under uniaxial tension, (**b**) force-displacement curves under notch tension, and (**c**) force-displacement under punch test

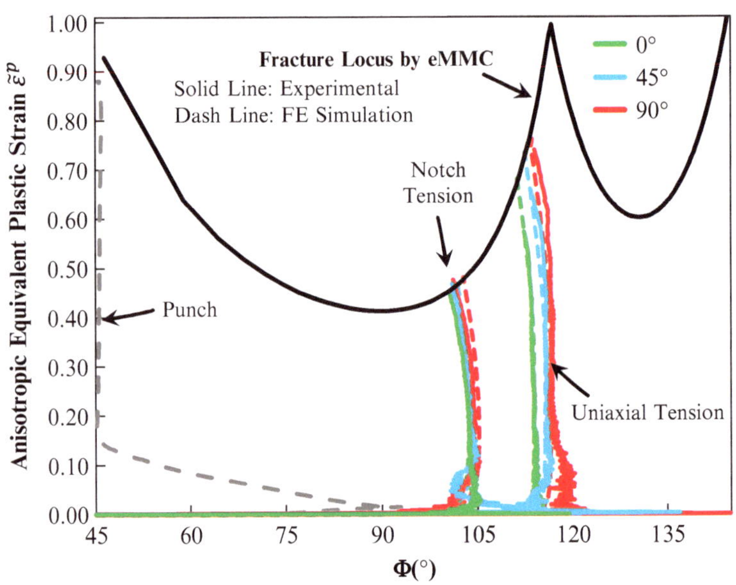


**Fig. 8.3** Comparison between tested and simulated strain histories for TRIP780, in the space of equivalent plastic strain versus strain angle Φ

**Table 8.1** Anisotropic all-strain based MMC coefficients for TRIP780

| | |
|---|---|
| $A$ | 1275.9 MPa |
| $n$ | 0.265 |
| $C_1$ | 0.121 |
| $C_2$ | 720.2 MPa |
| $\widetilde{C}_\theta^s$ | 1.0956 |
| $\widetilde{C}_\theta^c$ | 1.0008 |
| $\beta_{11}$ | 1.000 |
| $\beta_{22}$ | 1.000 |
| $\beta_{33}$ | 1.000 |
| $\beta_{12}$ | 1.000 |

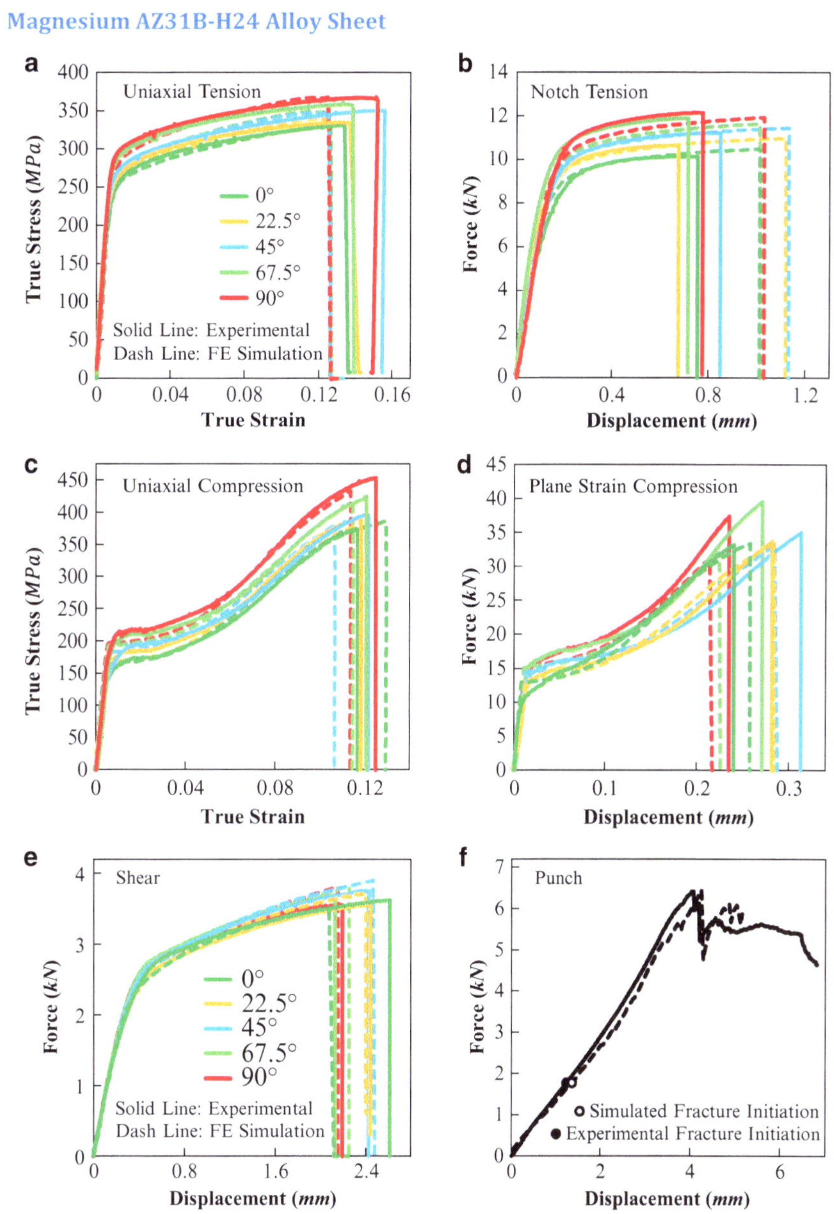


**Fig. 8.4** Comparison between tested and simulated results of AZ31B-H24, in (**a**) true stress-strain under uniaxial tension, (**b**) force-displacement under notch tension, (**c**) true stress-strain under uniaxial compression, (**d**) force-displacement under plane strain compression, (**e**) force-displacement under shear, and (**f**) force-displacement under punch

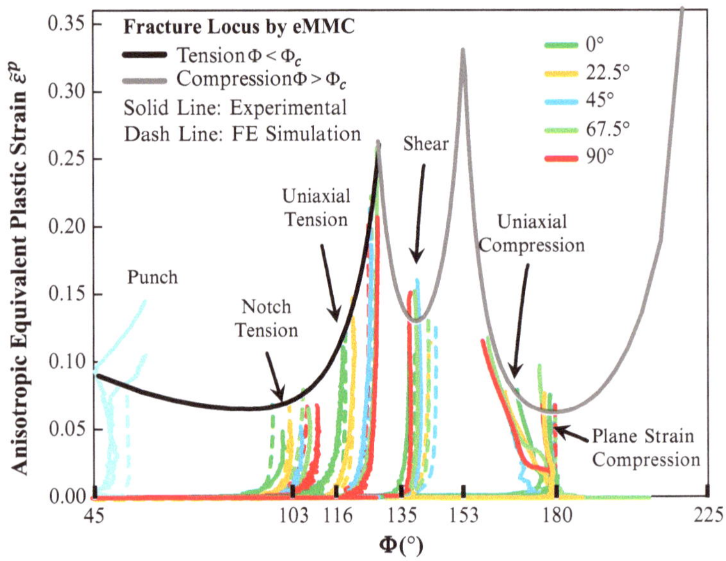


**Fig. 8.5** Comparison between tested and simulated strain histories for AZ31B-H24, in the space of equivalent plastic strain versus strain angle Φ

**Table 8.2** Anisotropic all-strain based MMC coefficients for magnesium AZ31B-H24

| | Φ Angle | |
|---|---|---|
| | $\leq \Phi_C$ | $> \Phi_C$ |
| $A$ | 409.4 MPa | |
| $n$ | 0.156 | |
| $C_1$ | 0.427 | −0.085 |
| $C_2$ | 217.6 MPa | |
| $\widetilde{C}_\theta^s$ | 0.973 | 1.214 |
| $\widetilde{C}_\theta^c$ | 1.030 | 1.246 |
| $\beta_{11}$ | 1.032 | |
| $\beta_{22}$ | 0.505 | |
| $\beta_{33}$ | 0.587 | |
| $\beta_{12}$ | 1.054 | |

## 8.4 Conclusions

The original stress invariants based MMC fracture model was transferred into an all-strain based MMC model under the plane stress condition, predicting the fracture strain in the space of strain ratio or Φ angle, instead of stress triaxiality and Lode angle parameter. The strain ratio or Φ angle can be directly measured by DIC. This approach makes it possible to study the fracture behavior of materials while bypassing plasticity for materials with complex plastic properties. The fracture strain was extended by a non-conjugated anisotropic equivalent strain with a linear transformation to incorporate the anisotropic fracture behavior. The eMMC fracture locus for TRIP780 steel sheets and magnesium AZ31B-H24 alloy sheets were incorporated into FE analysis, correlating very well with experimental results.

**Acknowledgments** Partial financial supports from Volkswagen (Germany), State Key Laboratory of Automotive Safety and Energy, Tsinghua University (Beijing, China), and University of Central Florida (UCF) are greatly appreciated.

## References

1. Bai, Y., Wierzbicki, T.: A new model of metal plasticity and fracture with pressure and Lode dependence. Int. J. Plast. **24**, 1071–1096 (2008)
2. Bai, Y.: Effect of loading history on necking and fracture. PhD thesis, Massachusetts Institute of Technology (2008)
3. Barsoum, I., Faleskog, J.: Rupture mechanisms in combined tension and shear–experiments. Int. J. Solids Struct. **44**(6), 1768–1786 (2007)
4. Bai, Y., Wierzbicki, T.: Application of extended Mohr-Coulomb criterion to ductile fracture. Int. J. Fract. **161**, 1–20 (2010)

5. Jia, Y., Long, X., Wang, K., Bai, Y.: Calibration of plasticity and fracture of magnesium alloy sheets under biaxial loading conditions. International Symposium on Plasticity and its Current Applications, Bahamas, 3–8 January (2013)
6. Bai, Y., Atkins, T.: Tension and shear cracking during indentation of ductile materials by opposed wedges. Eng. Fract. Mech. **96**, 49–60 (2012)
7. Luo, M., Dunand, M., Mohr, D.: Experiments and modeling of anisotropic aluminum extrusions under multi-axial loading, part II: Ductile fracture. Int. J. Plast. **32–33**, 36–58 (2012)

# Chapter 9
# Hydrolytic Degradation and Its Effect on Mechanical Properties of HFPE-II-52 Polyimide: Preliminary Results

**Yi Xu and Alan T. Zehnder**

**Abstract** Polyimides and fiber reinforced polyimide matrix composites are used in demanding applications requiring mechanical performance at high temperatures (300 + °C). When exposed to moisture and elevated temperature for extended periods of time polyimides may undergo hydrolytic degradation in which bonds are broken at a rate dependent on the temperature and moisture absorbed into the material. These broken bonds will be reflected in reductions in mechanical properties such as moduli, glass transition temperature and flow strength. In this project the polyimide HFPE-II-52 is aged under fully moisture saturated conditions at temperatures up to 250 °C. Following the temperature and moisture exposure, cube shaped samples are tested in compression at room and elevated temperatures to measure the reductions in stiffness and yield stress. This research is aimed at providing a means for monitoring and predicting hydrolytic degradation and its effect on mechanical performance.

**Keywords** Polyimide • Hydrolysis • Damage • High temperature • Mechanical behavior

## 9.1 Introduction

Polyimide matrix composites (PiMCs) extend the role of traditional polymer matrix composites (PMCs) to high temperatures due to their high glass transition temperature ($T_g$ around 350 °C), toughness and stiffness. However, polyimides or PiMCs, when exposed to hot, high humidity environments, can suffer significant degradation and can fail by mechanisms such as blistering and delamination [1, 2]. One of the underlying mechanisms of moisture degradation is hydrolysis, or breaking of bonds due to reaction with water. Hydrolysis can be reflected in reductions of mechanical properties such as elastic moduli and flow strength. In early work on this topic the tensile strength of Kapton polyimide film immersed in water at temperatures from 20 to 100 °C was shown to decrease as a function of immersion time [3]. In subsequent studies the degradation temperature was extended beyond 100 °C by using a heated, sealed pressure chamber partially filled with water. When heated, the environment equilibrates to at the steam saturation pressure. Studies of the time-temperature threshold of hydrolysis have been performed for the polyimides AFR700B, PETI-5, K3B and others [4, 5].

In this study, we apply the pressure chamber method to investigate the hydrolytic degradation of the polyimide HFPE-II-52. We expose the polyimide samples to fully saturated conditions at temperatures up to 250 °C, and for time spans from several hours to a week. The exposed samples are then tested in compression at both room and high temperature (285 °C) to probe the resulting changes in moduli and strength. This is a work in progress, but preliminary results are given here and discussed.

## 9.2 Experimental Method

### *9.2.1 Sample Preparation*

The samples used for hydrolytic degradation tests were manufactured by compression molding of HFPE-II-52 polyimide powder at 172 atm pressure and at a maximum temperature of 377 °C for 2 h. The polyimide plates were then postcured at 371 °C for 16 h. Details on HFPE-II-52 synthesis, molding procedure and cure cycle can be found in reference [6].

Y. Xu • A.T. Zehnder (✉)
Field of Theoretical and Applied Mechanics, Cornell University, Ithaca, NY 14853, USA
e-mail: atz2@cornell.edu



A.T. Zehnder et al. (eds.), *Fracture, Fatigue, Failure and Damage Evolution, Volume 8*,
Conference Proceedings of the Society for Experimental Mechanics Series, DOI 10.1007/978-3-319-42195-7_9

The samples were then cut into 5 × 5 × 4 mm specimens from a 100 × 100 × 4 mm plate with a diamond wire saw. Before moisture exposure and testing all samples were dried for 3 days at 70 °C in a vacuum chamber.

### 9.2.2 *Degradation Method*

We use a capped, threaded, stainless steel tube to create an environment that can hold saturated steam pressure at temperatures up to 300 °C, see Fig. 9.1. Thread tape is used to help maintain the seal and to minimize seizing of the cap to the tube. In the tube, the polyimide specimen sits on a piece of stainless steel mesh. Underneath the mesh, a pack of fiber glass soaks enough water so that the steam will be saturated once the sample reaches the desired degradation temperature. This set up prevents the sample from directly interacting with liquid water, since it is suspected that the hydrolysis process may be different with exposure to liquid water rather than steam [4]. Distilled water is used to minimize chemical reactions other than hydrolysis. The tube is then heated in a chamber to the desired temperatures. After the set exposure time, the chamber is allowed to cool, the sample removed and again dried in a vacuum chamber. All samples are weighed at three time points—prior to moisture exposure, right after moisture exposure and after being dried prior to compression testing.

### 9.2.3 *Compression Tests*

The set up for the compression tests is shown in Fig. 9.2. The experiments are performed using an MTS 45 kN servo-hydraulic testing system. Two cylindrical stainless steel rods are used to compress the sample. Both room temperature and high temperature (285 °C) tests are conducted. For high temperature tests, a clam shell furnace ATS 3210 is used to heat the sample. Two K-type thermocouples are mounted in different places of the upper rod to control the temperature. The tips of the thermocouples are placed just underneath the upper rod. The specimen is heated until the average of the two thermocouples indicates a temperature of 285 °C. An additional hour is allowed for the system to reach thermal equilibrium. To make sure that the center of sample indeed reaches to the desired temperature during the test, a dummy specimen was

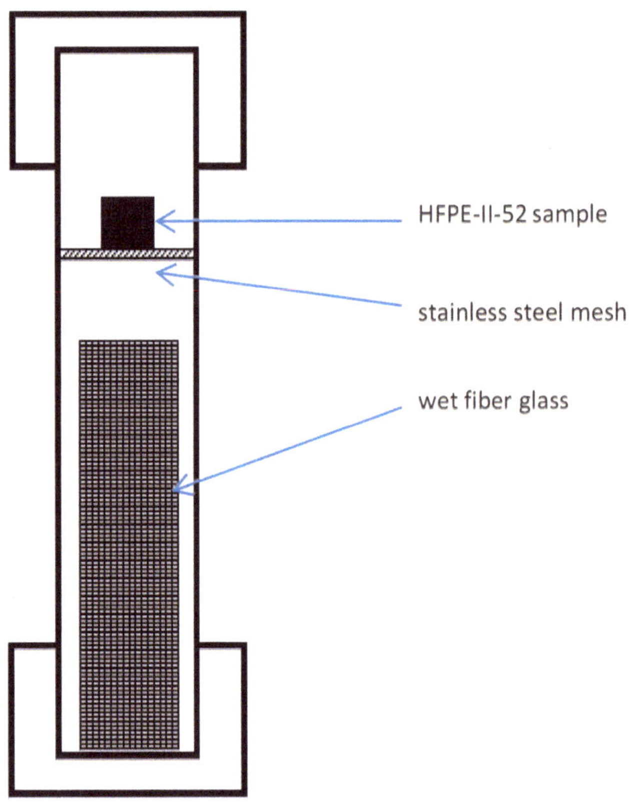


**Fig. 9.1** Sealed stainless steel tube set-up

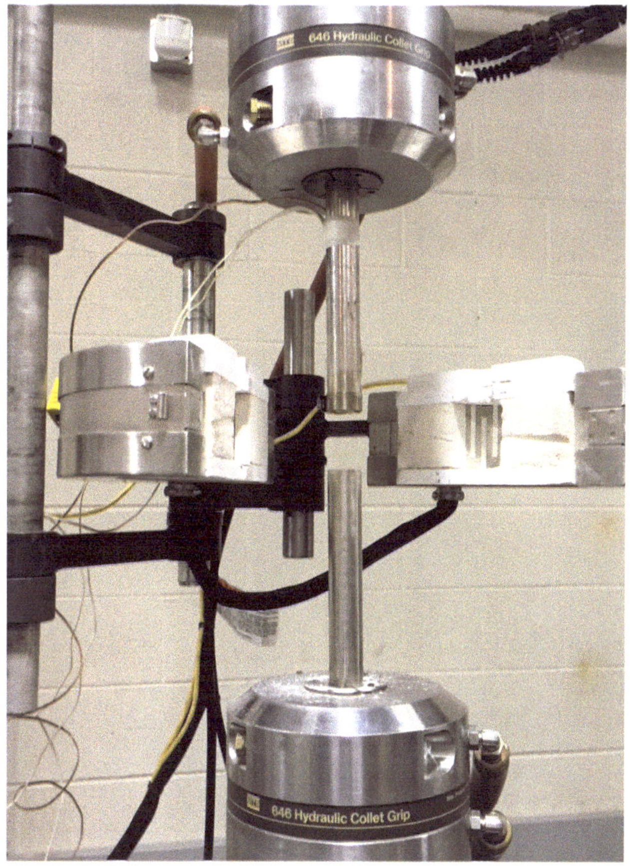


**Fig. 9.2** High temperature compression test set-up

created by gluing two samples together with a thermocouple placed between them. We have found that the difference between the temperature at the center of dummy specimen and the average of the temperature indicated by thermocouples used for controlling the temperature is smaller than 3 °C. For both room and high temperature tests, the samples are loaded until failure with displacement control at a strain rate of approximately $10^{-3}$/s. The compressive stress-strain curve is then obtained for analysis of material modulus and strength reductions due to hydrolytic degradation.

## 9.3 Preliminary Results

### *9.3.1 Weight and Visible Changes*

The maximum amount of water that HFPE-II-52 polyimide can absorb under 1 atm is 3.2 % [7]. The maximum weight gain is independent of temperature when the saturation experiment is performed below 100 °C. However, we have found a higher amount of weight gain for polyimide under saturated steam pressure at high temperatures. For moisture exposures at temperatures from 125 to 200 °C, and up to 3 days, a weight gain of approximately 3.5 % is recorded. No significant visible changes are observed on the samples' surface and after drying no significant (<0.5 %) weight lost is found compared to the weight before moisture saturation. However at 225 °C an average of 42 % weight gain is found for exposure of only 1 day. We also observe a large swelling strain of around 20 %. The surface of the specimen wrinkles and starts to turn yellow, which indicates a change of structure of polyimide during the hydrolysis process. After the sample has been dried, an average 3.7 % weight lost is found.

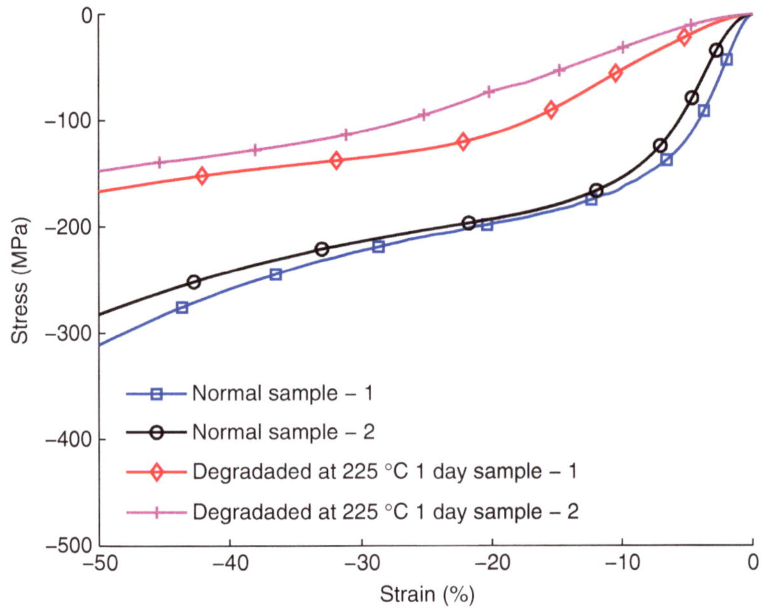


**Fig. 9.3** Compressive stress strain curve testing at room temperature. True stress and true strain measurements. The markers in the plots identify the curves only; they do not signify individual data points

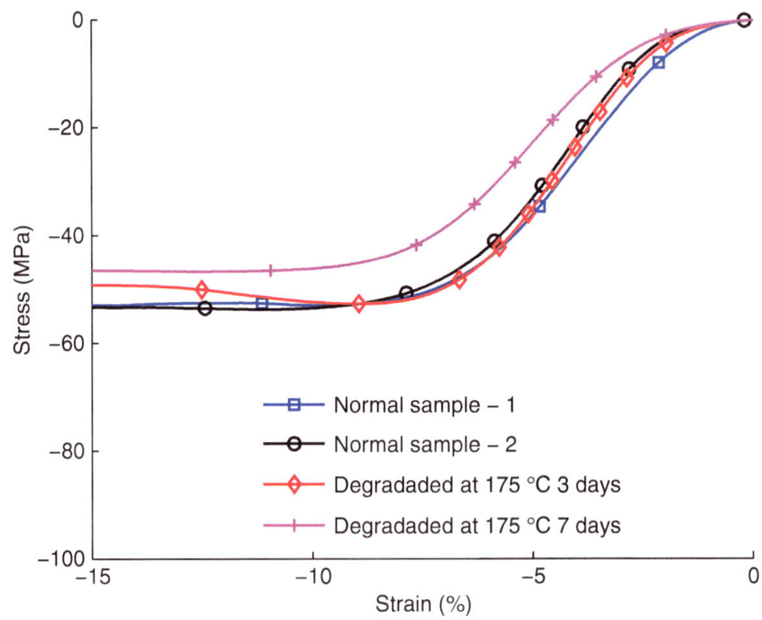


**Fig. 9.4** Compressive stress strain curve testing at 285 °C. True stress and true strain measurements. The markers in the plots identify the curves only; they do not signify individual data points

### *9.3.2 Compression Test Results*

Sample data from the compression are shown in Figs. 9.3 and 9.4. For room temperature compression tests, no significant changes are obtained at degradation temperatures below or equal to 200 °C, up to 7 days exposure. For 225 °C 1 day exposure, the material retains only 22 % of its initial modulus and 76 % of the initial yield stress (determined by the 0.2 % offset method). The degradation effects, however, are magnified for compression tests performed at high temperature (285 °C). Reductions of both material moduli (81 % remaining) and yield stress (88 % remaining) are observed for 175 °C, 7 days exposure.

## 9.4 Conclusion and Future Work

A hydrolytic degradation experiment at high temperature has been conducted using a sealed stainless steel tube to provide the high temperature saturated steam environment. Preliminary results show that the hydrolysis of HFPE-II-52 polyimide is time and temperature dependent. More hydrolysable bonds break at higher temperatures and with longer duration exposures to moisture at high temperature. The hydrolysis effect is reflected in reductions of stiffness and yield stress as measured in a compression test. Future work will involve refinement of the test methods, additional tests to obtain more conclusive data and use of the data to develop a constitutive model that incorporates the time, temperature and moisture dependent hydrolytic degradation of the polyimide. Such a model will be of value in the predication of performance and life-time of polyimide and PiMC components.

## References

1. Czabaj, M.W., Zehnder, A.T., Chuang, K.: Blistering of moisture saturated graphite/polyimide composites due to rapid heating. J. Compos. Mater. **43**(2), 153–174 (2008)
2. Czabaj, M.W., Zehnder, A.T., Hui, C.: Delamination of moisture saturated graphite/polyimide composites due to rapid heating. Compos. Part B **41**(7), 568–577 (2010)
3. Delasi, R., Russell, J.: Aqueous degradation of polyimides. J. Appl. Polym. Sci. **15**(12), 2965–2974 (1971)
4. Shin, E.E., Morgan, R.J., Zhou, J., Sutter, J.K., Meador, M.A.: High temperature polymer matrix-carbon fiber composites-critical degradation mechanisms and test methodologies. In: 44th International SAMPE Symposium and Exhibition, Long Beach, CA, pp. 2382–2396, 23–27 May 1999
5. Shin, E.E., Morgan, R.J., Zhou, J.: Hydrolytic degradation mechanisms and kinetics of polyimides for advanced composites. In: 45th International SAMPE Symposium and Exhibition, Long Beach, CA, p. 389 21–25 May 2000
6. Antonakakis, J.: Linear viscoelastic properties of HFPE-II-52 polyimide. M.S. Thesis, Cornell University, Ithaca, NY (2005)
7. Bhargava, P., Zehnder, A.T.: Moisture diffusion properties of HFPE-II-52 polyimide. J. Appl. Polym. Sci. **102**(4), 3471–3479 (2006)

# Chapter 10
# Mixed-Mode and Mode-II Fatigue Crack Growth in Woven Composites

**Joel S. Fenner and Isaac M. Daniel**

**Abstract** A woven carbon/epoxy composite was subjected to fatigue crack growth under mixed Mode-I/Mode-II loading to obtain crack growth behavior at different cyclic strain energies. Owing to the woven structure of the material, pure Mode-II fracture is usually a difficult proposition because of friction, interference, and interlock of woven tows in adjacent plies at an interlaminar crack. These limitations were overcome by the use of a novel form of mixed Mode-I/Mode-II specimen, which imposes sufficient crack surface opening (Mode-I) to alleviate ply–ply interactions, but not so much as to obscure the sliding (Mode-II) response. Comparison with pure Mode-I fatigue crack growth data, in conjunction with a fracture interaction criterion, provided a means to extract the Mode-II behavior.

**Keywords** Composites • Fracture • Fatigue • Mixed-Mode • Test Methods

## 10.1 Introduction

Composites comprised of woven fibers embody a significant portion of commercial composite material production, owing to the high performance of finished parts and the relative ease of handling the raw materials (preforms, prepreg). However, in spite of their unquestionable practicality, the additional geometric complexity introduced by the woven structure of the fibers (crimps) in such composites has many confounding consequences on the characterization and testing of these materials.

In the case of fracture, the complexity of the woven structure has different effects depending on the mode of crack propagation. For instance, in the case of Mode-I crack growth (opening mode), the propagation of an interlaminar crack around the crimps of the woven structure has an effect of introducing significant tortuosity to the crack path, and this often will manifest itself as a 'stepwise' or 'zig–zag' load-displacement character in comparison to a more uniform unidirectional material (Fig. 10.1) that lacks the more complicated geometry of the fiber arrangement. Such effects can often be accommodated reasonably well in data analysis by averaging methods which look at overall material response rather than local behavior at any given instant of crack propagation around a feature of the woven structure [1–3].

In the case of a Mode-II process (sliding mode), the situations is, however, quite worse. One of the more significant challenges in Mode-II fracture testing of woven composites is the 'interlock' phenomenon associated with the woven structure. In essence, the crimps of the fabric weave, in adjacent plies, have a strong tendency to physically interfere with one another, opposing sliding of one ply against another. Structurally, this is beneficial, as it presents a mechanism of resistance to the sliding motion required for extension of a crack in Mode-II (sliding mode) that is more robust than in unidirectional composites (Fig. 10.2). From a materials testing standpoint, however, it is quite confounding, as it prevents the successful execution of many forms of convenient Mode-II fracture test because the nascent crack fails to propagate properly beyond initial release, and initial release is sometimes suspect.

Because Mode-I crack propagation (opening mode) is not so fundamentally opposed or thwarted by the more complicated path a crack must take in a woven composite, this effect of ply-on-ply interlock can generally be ignored or treated as irrelevant in data analysis. It can then readily be seen how imposing some amount of Mode-I opening onto an otherwise "pure" Mode-II sliding test might give a means of alleviating the interlock effect while still preserving enough of the Mode-II behavior of a test to measure, which is the principal aim of this research.

J.S. Fenner (✉) • I.M. Daniel
Center for Intelligent Processing of Composites, Northwestern University, 2137 Tech Drive, Evanston, IL 60208, USA
e-mail: joelfenner2012@u.northwestern.edu

 
A.T. Zehnder et al. (eds.), *Fracture, Fatigue, Failure and Damage Evolution, Volume 8*,
Conference Proceedings of the Society for Experimental Mechanics Series, DOI 10.1007/978-3-319-42195-7_10

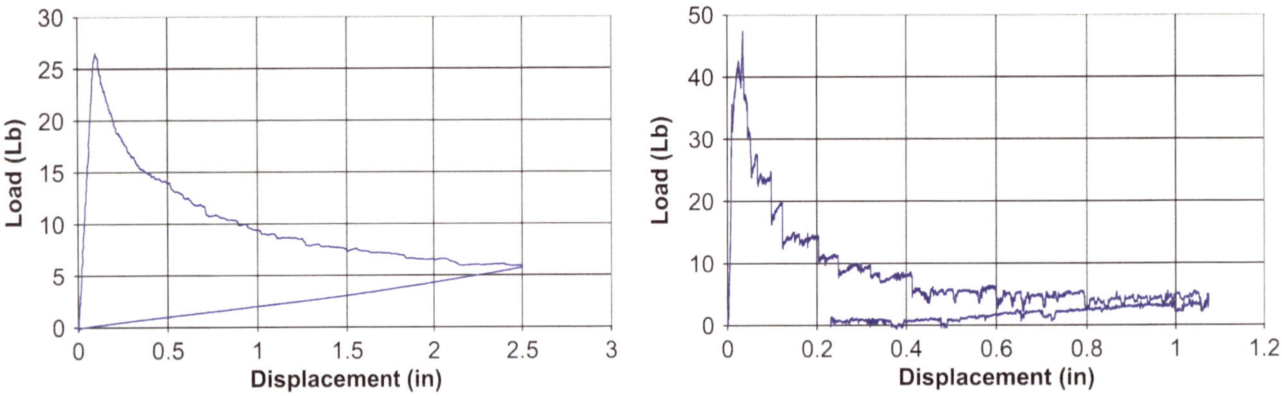


**Fig. 10.1** Representative load-displacement data from similar Mode-I fixed-width double cantilever beam (DCB) fracture specimens. *Left*: unidirectional carbon/epoxy composite, *Right*: woven carbon/epoxy composite, showing 'stepwise' behavior

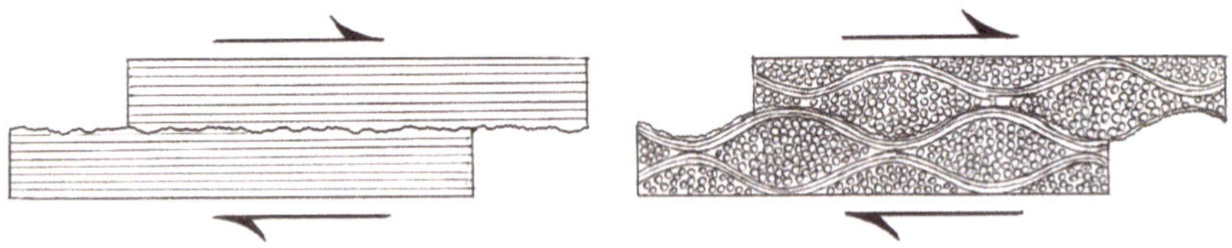

**Fig. 10.2** Effect of material structure on a Mode-II (sliding) interlaminar fracture process. *Left*: unidirectional material showing relatively uninhibited planar growth. *Right*: woven material showing effect of mechanical interference of crimps impeding movement of adjacent crack faces

## 10.2 Material Preparation

Two types of carbon/epoxy composite were employed in this study to achieve two different aims. A unidirectional composite material was prepared from prepreg of Aksaca A42 fibers with Dow P6300 IMR resin as the matrix material. Curing was performed in a heated hydraulic press to produce finished sheets of composite at a final fiber volume fraction of approximately 51 %. A woven composite material was prepared from dry 5-harness satin AS4 fiber preform (as Hexcel AGP370-5H) and an anhydride-cured epoxy (Huntsman GY6010-HY917-DY070 in weight proportions 100:90:1) in a wet layup process with subsequent vacuum debulking and curing in a heated hydraulic press. In both cases, the curing temperature was 150 °C. Composite sheets as cast were nominally 300 mm × 300 mm × 2.5 mm. Samples were abrasively machined from these sheets using diamond abrasive cutting wheels and SiC abrasive papers to achieve desired geometries.

## 10.3 Mode-I Testing

To calibrate the response of the mixed-mode specimen at the heart of the study, pure single-mode characterization tests were first carried out on the unidirectional material, as its simpler microstructure typically gives experimental data that are easier to interpret than woven material. The Mode-I critical strain energy release rate ($G_{Ic}$) was obtained by testing of width-tapered DCB specimens [4] (Fig. 10.3) under quasi-static conditions, obtaining load-displacement curves as in Fig. 10.4 (left).

For the width-tapered DCB specimen, the specimen compliance is found from Timoshenko beam theory to be

$$C = \frac{\delta}{P} = \frac{12}{E_1 h^3}\left[\left(a^2 - s^2\right)k + \frac{2s}{3B}\right] + \frac{12}{5hG_{13}}\left[\ln\left(\frac{a}{s}\right) + \frac{s}{B}\right] \tag{10.1}$$

where $\delta$ is the end deflection of the specimen
$P$ is the applied load
$E_1$ is the longitudinal elastic modulus of the composite material

**Fig. 10.3** Illustration of typical width-tapered DCB Mode-I fracture specimen

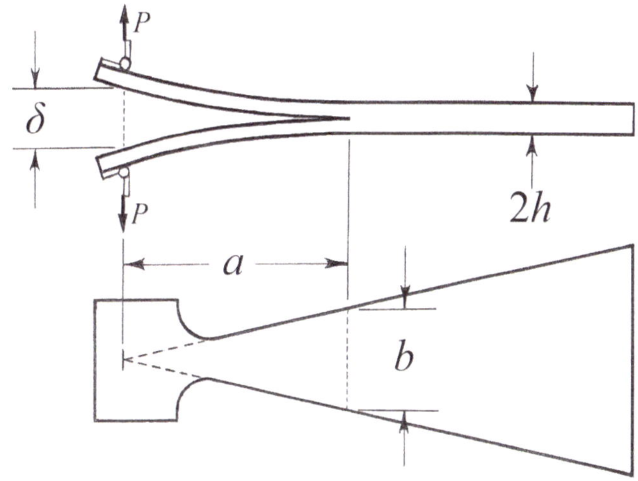


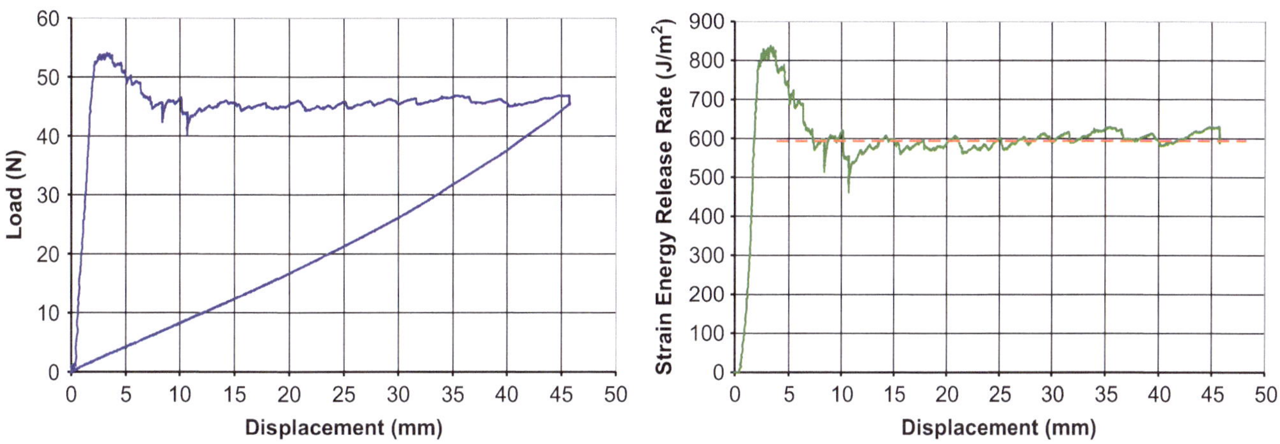


**Fig. 10.4** Representative curves obtained from Mode-I static fracture testing of unidirectional composite. *Left*: Load-displacement curve. *Right*: Plot of instantaneous strain energy release rate $G_I$ during test

$G_{13}$ is the shear modulus of the composite material in plane of bending
$a$ is the crack length
$s$ is the tab length
$B$ is the tab width
$k = a/b$ is the width taper ratio
2 $h$ is the specimen thickness

Because testing is conducted in a uniaxial manner, the compliance method may be applied to obtain the Mode-I strain energy release rate $G_I$ at any given instant as

$$G_I = \frac{1}{2}\frac{P^2}{b}\frac{dC}{da} = \frac{12P^2k}{E_1ha^2}\left[k\left(\frac{a}{h}\right)^2 + \frac{1}{10}\left(\frac{E_1}{G_{13}}\right)\right] \tag{10.2}$$

It is readily apparent from this expression that, for conditions where the sample aspect ratio is high ($a >> h$) or where shear effects may otherwise be neglected (e.g. as $G_{13} \to \infty$), this reduces to the expression

$$G_I \cong \frac{12P^2k^2}{E_1h^3} \tag{10.3}$$

It is worth noting that this expression does not contain the crack length $a$ and that the width-tapered DCB specimen is thus generally insensitive to the crack length, requiring no effort to measure it in order to obtain an accurate value for the critical

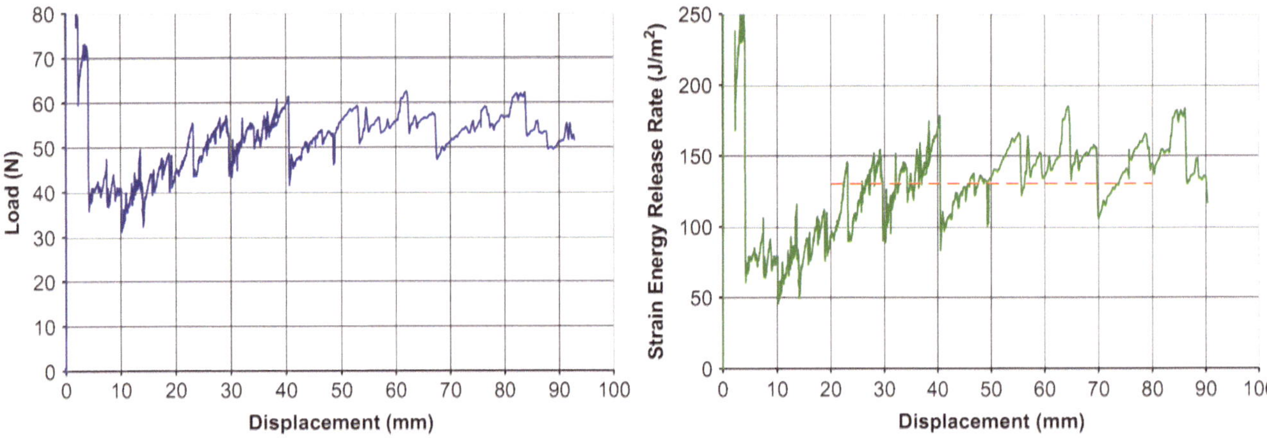


**Fig. 10.5** Representative curves obtained from Mode-I static fracture testing of woven composite. *Left*: Load-displacement curve. *Right*: Plot of instantaneous strain energy release rate $G_I$ during test

**Fig. 10.6** Illustration of typical fixed-width edge notched flexure (ENF) Mode-II fracture specimen

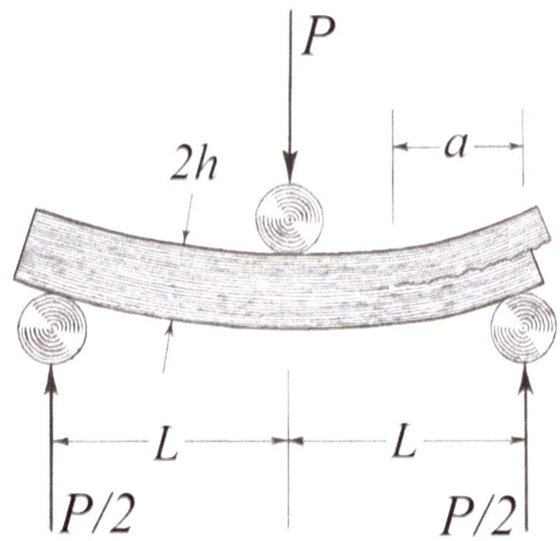


strain energy release rate $G_{Ic}$. Applying Eq. (10.3) to the load displacement data gives a plot of the strain energy release rate as a function of sample displacement, showing a relatively constant character.

Similar tests were subsequently performed on woven material, giving load-displacement curves and strain energy release rate response as in Fig. 10.5.

## 10.4 Mode-II Testing

Pure Mode-II static fracture tests were conducted using the ubiquitous edge notched flexure (ENF) specimen [5] to provide the remaining necessary information to calibrate the mixed-mode response. The Mode-II critical strain energy release rate ($G_{IIc}$) was obtained by testing of these specimens (Fig. 10.6) under quasi-static conditions, obtaining load-displacement curves as in Fig. 10.7.

For the ENF specimen, the specimen compliance is found from Timoshenko beam theory to be

$$C = \frac{\delta}{P} = \frac{1}{8bh}\left[\frac{2L^3 + 3a^2}{E_1 h^2} + \frac{12L + 9a}{5G_{13}}\right] \tag{10.4}$$

where $L$ is the half-span of the ENF specimen

$b$ is the specimen width

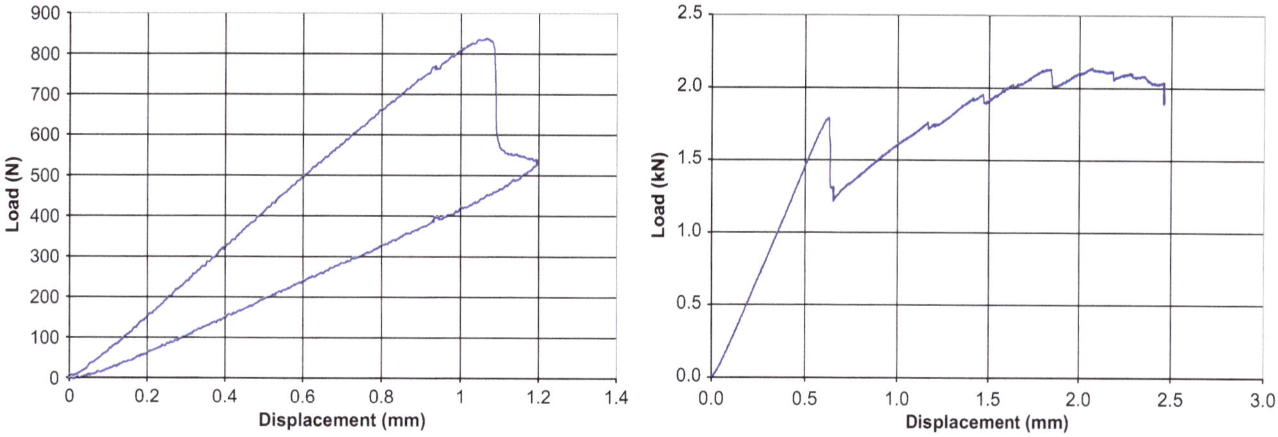


**Fig. 10.7** Representative load-displacement plots from Mode-II ENF static test. *Left*: unidirectional composite; *Right*: woven composite

**Fig. 10.8** Illustration of idealized ENF specimen load-displacement response and associated physical phenomena

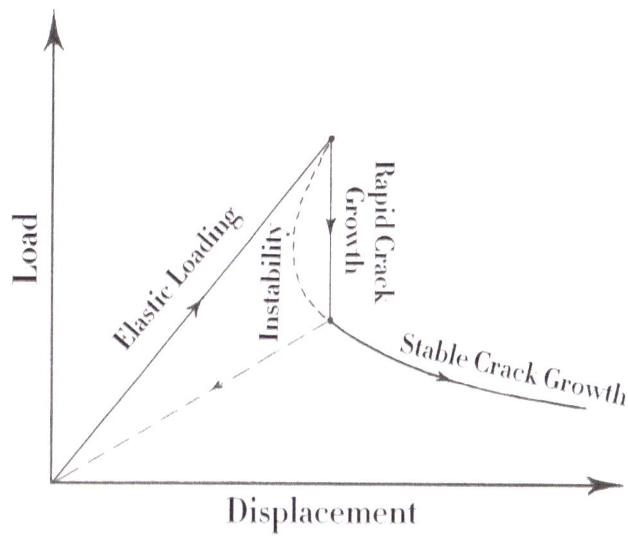


The compliance method is again readily invoked to obtain the Mode-II strain energy release rate of

$$G_{II} = \frac{1}{2}\frac{P^2}{b}\frac{dC}{da} = \frac{9P^2}{16E_1bh}\left[\left(\frac{a}{h}\right)^2 + \frac{1}{5}\left(\frac{E_1}{G_{13}}\right)\right] \tag{10.5}$$

Unlike the Mode-I DCB, the Mode-II ENF generally lacks a stable crack growth character for initial crack lengths shorter than $a \cong 0.7L$ [5]. Hence, only the initial release is of significance, and one may only rely on the peak load prior to release to obtain the Mode-II critical strain energy release rate. The expected load-displacement character (Fig. 10.8) is one in which there exists a steep drop in load at first release, followed by a monotonically declining load as the crack is extended further. It is noteworthy that the unidirectional material largely exhibits this physical behavior (Fig. 10.7 left—a small portion of the declining load after peak can be seen) in the load-displacement curves as one would expect, but the woven material does not. In fact, the load drop at release in the woven material is comparatively small, and if a woven sample is deformed beyond this point, it exhibits an increasing load characteristic (Fig. 10.7 right) that does not agree with the expected declining load and stable crack growth response obtained from theory. In essence, this is a consequence of the ply-on-ply interlock which is the confounding effect at the heart of many problems in Mode-II fracture testing of woven composites.

## 10.5 Mixed-Mode Testing

For the mixed-mode static tests (and subsequent fatigue crack growth tests), a somewhat novel specimen geometry was employed (Fig. 10.9). Conceptually, the specimen resembles its pure Mode-I width-tapered DCB cousin (as in Fig. 10.3), but with a few modifications to introduce a mixture of Mode-I and Mode-II. The method of imposing multiple modes is most readily understood by decomposing the end loading conditions into two more obvious cases of pure Mode-I and pure Mode-II behavior (Fig. 10.10).

The compliance of the physical specimen (mixed mode) is found from Timoshenko beam theory to be

$$C = \frac{\delta}{P} = \frac{k}{2\kappa h G_{13}}\left[1 + \ln\left(\frac{L^2}{s(L-a)}\right)\right] + \frac{k}{4E_1h^3}\left(3L_2 + 42La - 21a^2 - 3s^2\right) + \frac{2s^3}{4E_1Bh^3} + \frac{s}{2\kappa BhG_{13}} \tag{10.6}$$

where $\delta$ is the end deflection of the specimen
$P$ is the applied load at the end of the specimen
$L$ is the span between the fixed and simple loading points
$a$ is the length of the crack
$B$ is the width of the loading tab
$s$ is the length of the loading tab
$2h$ is the total specimen thickness
$\kappa$ is the Timoshenko shear coefficient ($\kappa = 5/6$ for rectangular cross section)

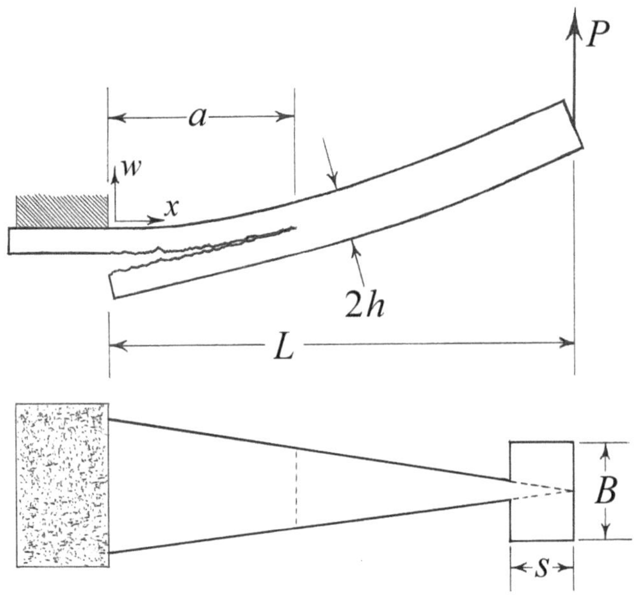


**Fig. 10.9** Illustration of mixed-mode fracture specimen

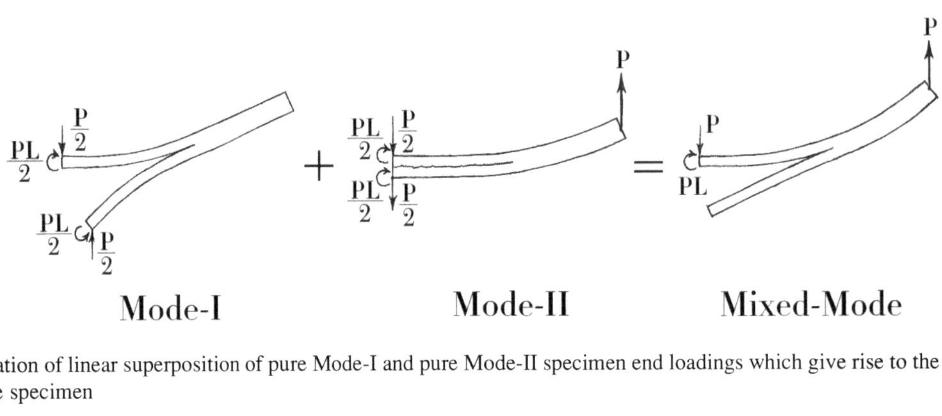


**Fig. 10.10** Illustration of linear superposition of pure Mode-I and pure Mode-II specimen end loadings which give rise to the loading state found in the mixed-mode specimen

$G_{13}$ is the shear modulus
$E_1$ is the longitudinal material elastic modulus
Applying the compliance method gives a total strain energy release rate of

$$G_T = \frac{P^2 k^2}{4h}\left[\frac{6}{5G}\left(\frac{1}{L-a}\right)^2 + \frac{21}{Eh^2}\right] \tag{10.7}$$

which, under conditions of negligible shear effects may be approximated as

$$G_T \cong \frac{21P^2k^2}{4E_1h^3} \tag{10.8}$$

A similar analysis is readily performed on the Mode-I and Mode-II component cases (as in Fig. 10.9), giving Mode-I and Mode-II component strain energy release rates of

$$G_I \cong \frac{3P^2k^2}{E_1h^3} \tag{10.9}$$

and

$$G_{II} \cong \frac{9P^2k^2}{4E_1h^3} \tag{10.10}$$

From this analysis, it is possible to then compute the ratio of the Mode-I strain energy release rate component to the total & the Mode-II strain energy release rate component to the total, giving

$$\frac{G_I}{G_T} = \frac{4}{7} \tag{10.11}$$

and

$$\frac{G_{II}}{G_T} = \frac{3}{7} \tag{10.12}$$

showing this specimen configuration to possess a constant proportion of Mode-I and Mode-II components regardless of crack length.

There are several sample geometries which may accomplish a similar kind of effect of imposing multiple modes [6–8], but the one presented here was selected because of attractive features which make it unique. In essence, the sample mimics the geometry of its pure Mode-I width-tapered DCB cousin as a means of obviating the need to directly measure the crack length. It also contains an inherent and constant ratio of fracture mode energies as the crack progresses, which is useful when conducting a test to interrogate the interaction of modes, as the test is uniform in its mode-mixing behavior from beginning to end.

The largest deficiency of this attack is the need to ultimately apply some form of mode mixing criterion to explain the interaction of the Mode-I and Mode-II energy release processes during cracking. There exist several models which can accomplish this [9–11], but unfortunately they are all semi-empirical in nature. The theory of Benzeggagh and Kenane was chosen for this work because of prior experimental efforts that had shown its applicability [12].

In its original form, the Benzeggagh–Kenane semi-empirical relation describing fracture mode interaction is

$$G_T = G_I + (G_{II} - G_I)\left(\frac{G_{II}}{G_T}\right)^m \tag{10.13}$$

where $G_T$ is the total specimen (mixed mode) strain energy release rate
$G_I$ is the Mode-I strain energy release rate component

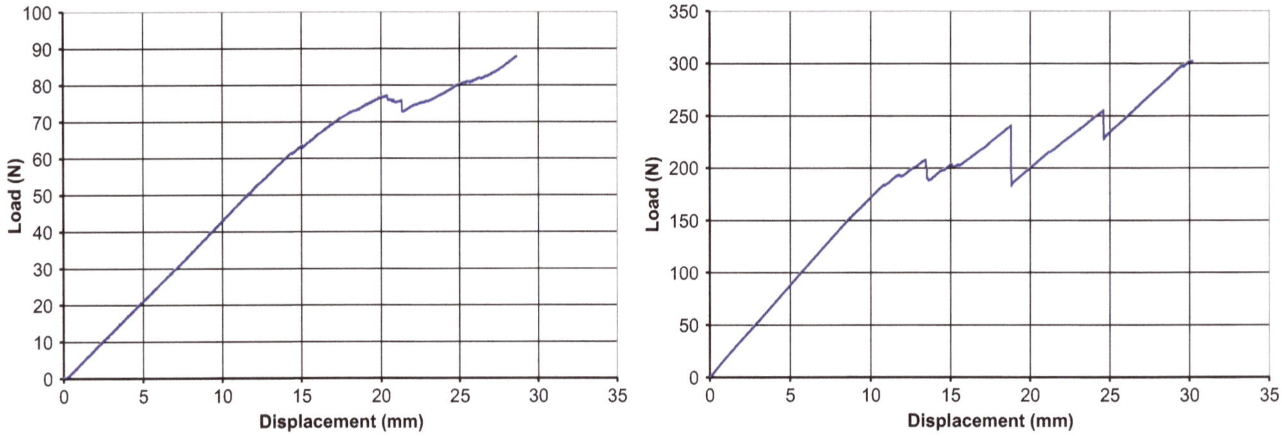


**Fig. 10.11** Representative mixed-mode specimen load-displacement curves. *Left*: unidirectional composite; *Right*: woven composite

**Table 10.1** Measured strain energy release rates for unidirectional carbon/epoxy composite

| Property | Value (J/m$^2$) | CoV (%) |
|---|---|---|
| Mode-I ($G_{Ic}$) | 537 | 12.8 |
| Mode-II ($G_{Ic}$) | 913 | 17.6 |
| Mixed-Mode ($G_T$) | 1100 | 29.1 |

**Table 10.2** Measured strain energy release rates for woven carbon/epoxy composite

| Property | Value (J/m$^2$) | CoV (%) |
|---|---|---|
| Mode-I ($G_{Ic}$) | 163 | 19.0 |
| Mode-II ($G_{Ic}$) | 590 | 30.2 |
| Mixed-Mode ($G_T$) | 2020 | 33.4 |

$G_{II}$ is the Mode-II strain energy release rate component
$m$ is an empirical interaction parameter

Because the ratio between the Mode-II strain energy release rate component $G_I$ and the total strain energy release rate $G_T$ are essentially fixed (Eqs. (10.11) and (10.12)), this relation collapses to a simple linear rule-of-mixtures expression as

$$G_T = G_I\left[1 - \left(\frac{3}{7}\right)^m\right] + G_{II}\left(\frac{3}{7}\right)^m \tag{10.14}$$

Hence, given measurements for $G_{Ic}$, $G_{IIc}$, and $G_{Tc}$ by experiment, the mode mixing parameter is obtained as

$$m = -\frac{\log\left(\frac{G_{Tc}-G_{Ic}}{G_{IIc}-G_{Ic}}\right)}{\log\left(\frac{3}{7}\right)} \tag{10.15}$$

Samples of the mixed mode type were then tested under displacement-controlled conditions, giving load-displacement curves as in Fig. 10.11. Ideally, these specimens should have given a constant plateau-load behavior upon onset of crack growth, as in the Mode-I width-tapered DCB tests (Fig. 10.5). However, there was a general trend toward monotonically increasing load in all samples tested, and the woven samples exhibited much more severe 'zig–zag' or 'stepwise' jumps than in Mode-I tests.

Collecting the data from the pure Mode-I width-tapered DCB tests, the pure Mode-II ENF tests, and the mixed-mode tests, the critical strain energy release rates for the two materials were found to be as follows.

Application of the data in Tables 10.1 and 10.2 to Eq. (10.15) then gave Benzeggagh–Kenane interaction parameters of
$m = 0.476$ for the unidirectional composite
$m = 1.735$ for the woven composite

This result is not altogether surprising, because the matrix behavior of the two materials is rather different, with the unidirectional material possessing a much less brittle matrix than the woven material. Furthermore, the result for the woven material is generally in agreement with Benzeggagh and Kenane's assertion that $m = 2$ is generally a good value for most brittle materials [10].

## 10.6 Mixed-Mode Fatigue Crack Growth

An interesting feature of the mixed mode specimen (and of some other width-tapered specimens [13]) is the behavior under conditions of an applied cyclic load $\Delta P$ permits continuous testing at a constant crack growth rate. In describing fatigue crack growth in a brittle material, it is often convenient to invoke the semi-empirical Paris law to describe the relationship between applied cyclic strain energy and rate of crack extension per cycle. The Paris law may be stated as

$$\frac{da}{dN} = C(\Delta K)^n \tag{10.16}$$

where $a$ is the crack length

$N$ is the number of cycles

$\Delta K$ is the cyclic applied stress intensity

$C$ and $n$ are empirically determined constants

Rearranging the Benzeggagh–Kenane relation using the expressions for the Mode-I strain energy release rate component $G_I$ (Eq. (10.9)) and the total strain energy release rate $G_T$ (Eq. (10.8)) of the mixed mode specimen gives an expression for the Mode-II component alone as

$$\frac{P^2k^2}{E_1h^3}\left[\frac{9}{4}\left(\frac{7}{3}\right)^m + 3\right] = G_{II} \tag{10.17}$$

Because beam bending specimens generally embody a case of an Eulerian beam, they may be treated as possessing a plane-stress case, and the relation between the stress intensity factors $K_i$ and strain energy release rates $G_i$ ($i = I,II,T$) is merely

$$\frac{K_i^2}{E_1} = G_i \tag{10.18}$$

If this is used to rewrite Eq. (10.17) in terms of the stress intensity factor $K_{II}$, the expression becomes

$$K_{II} = \frac{Pk}{h^{3/2}}\left[\frac{9}{4}\left(\frac{7}{3}\right)^m + 3\right]^{1/2}$$

If, however, a cyclic load is applied to the specimen ($\Delta P = P_{max} - P_{min}$), this expression simply becomes

$$\Delta K_{II} = (P_{max} - P_{min})\frac{k}{h^{3/2}}\left[\frac{9}{4}\left(\frac{7}{3}\right)^m + 3\right]^{1/2} \tag{10.19}$$

This expression can be substituted into the Paris Law relation (Eq. (10.16)), and thus a test conducted on a mixed-mode specimen under conditions of constant cyclic load amplitude $\Delta P$ should give data for a single crack growth rate da/dN related to the pure Mode-II response, giving a means of obtaining Mode-II cyclic crack growth behavior. This is significant because the mixed-mode specimen provides a means of alleviating the composite ply 'interlock' phenomenon that often obscures the true Mode-II fatigue crack growth behavior (Fig. 10.12).

Preliminary tests were then conducted using mixed-mode samples at different cyclic load amplitudes $\Delta P$ under tension-tension conditions with a load ratio $R = P_{min}/P_{max}$ of 0.1. The resultant measured specimen cyclic crack growth rates da/dN were then used to calculate the Paris Law parameters for the extracted Mode-II response using Eq. (10.19). The result of these calculations, as well as pure Mode-I Paris Law parameters obtained from prior work using only width-tapered DCB specimens (for comparison purposes) on the same material [14], are given in Table 10.3.

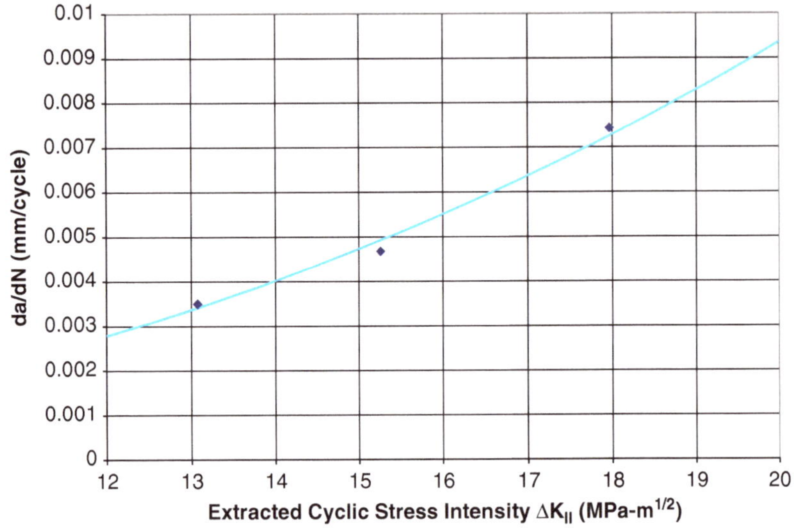


**Fig. 10.12** Extracted Mode-II fatigue crack growth behavior from mixed-mode testing and Paris law fit

**Table 10.3** Woven composite parameters for Paris law

| Mode | C | n |
|---|---|---|
| I | $2.049 \times 10^{-7}$ | 10.55 |
| II | $7.720 \times 10^{-6}$ | 2.37 |

Mode-II values extracted from mixed-mode sample behavior. Mode-I values from simple width-tapered DCB tests [14] on same material. N.B. Parameters $C$ and $n$ are for $\Delta K$ in MPa-m$^{1/2}$ and da/dN in mm/cycle

## 10.7 Conclusions

While the woven structure of some composites presents a unique challenge in the area of fracture testing, especially in cases of sliding-mode crack propagation, this research shows that it may be possible to alleviate some of these complications through the technique of mode mixing in a fracture test. The technique requires a greater amount of testing effort to obtain data for the pure Mode-I and Mode-II response of the material, which is not necessarily trivial to obtain, and also requires the use of a mixed-mode interaction criterion which is not as fundamental as other phenomena involved in fracture problems,. However, the benefit of utilizing a mixed-mode test is readily seen in situations where the woven structure of the material renders more conventional methods of fatigue crack growth testing ineffective.

The non-ideal response of the mixed-mode specimen, with monotonic increase in load over the full breadth of its range, implies some deviation from the fracture mechanics model, which is still cause for concern. Most likely, there is a complication arising from the fact that the available energy in the sample to create new crack surface (the strain energy associated with bending of the uncracked section) is constantly in decline as the crack progresses. One would expect that this would not matter, as the mixed-mode sample is width tapered and presents a smaller crack width as the crack progresses as well—the two phenomena should be in balance with one another. However, this basic process runs in opposition to samples like the Mode-I width-tapered DCB where the available energy is constantly increasing, while the crack width is increasing, and where the viability of the specimen geometry has been observed. The final implication is that in well-behaved width-tapered samples where the energy and crack width simultaneously increase, the incremental elastic energy increase most likely overtakes the increase in crack width by some small amount due to other effects (e.g. nonlinear beam behavior at large deflections). If the direction of crack propagation were reversed, it might alleviate this difficulty, and render the mixed-mode sample more predictable in its behavior.

**Acknowledgment** The work described here was sponsored by the Department of Energy, with the Ford Motor Company as a prime contractor.

## References


1. Jar, P.-Y.B., Compston, P.: Comparsion of interlaminar fracture toughness in unidirectional and woven roving marine composites. Appl. Compos. Mater. **5**, 189–206 (1998)
2. Reis, P.N.B., Ferreira, J.A.M., Costa, J.D.M., Pereira, A.M.: Interlaminar fracture in woven carbon/epoxy laminates. Fract. Struct. Integr. **30**, 431–437 (2014)
3. Shindo, Y., Sato, K., Narita, F., Sanada, K.: Mode-II interlaminar fracture and damage evaluation of GFRP woven laminates at cryogenic temperatures using the 4ENF specimen. J. Compos. Mater. **42**(11), 1089–1101 (2008)
4. Brussat, T.R., Chin, S.T., Mostovoy, S.: Fracture Mechanics for Structural Adhesive Bonds, Phase II, AFML-TR-77-163. Wright Aeronautical Labs, Dayton, OH (1978)
5. Carlsson, L.A., Gillespie, J.W., Pipes, J.R., Pipes, R.B.: On the analysis and design of the end notched flexure (ENF) Specimen for Mode II Testing. J. Compos. Mater. **20**, 594 (1986)
6. Dahlen, C., Springer, G.S.: Delamination growth in composites under cyclic loads. J. Compos. Mater. **28**(8), 732–781 (1994)
7. Wilkins, D.J., Eisenmann, J.R., Camin, R.A., Margolis, W.S., Benson, R.A.: Characterizing delamination growth in graphite-epoxy. In: Reifsnider, D. (ed.) Damage in Composite Materials, ASTM STP 775, pp. 168–183. West Consohocken, PA: American Society for Testing and Materials (1982)
8. Crews Jr., J.S., Reeer, J.R.: A Mixed-Mode Bending Apparatus for Delamination Testing NASA Tech Memorandum 100662. Nasa-Langley Research Center, Hampton, VA (1988)
9. Wu, E.M., Reuter, R.C.J.: Crack extension in fiberglass reinforced plastics and a critical examination of the general fracture criterion. Theoretical and Applied Mechanics Report No. 275 (Bureau of Naval Weapons, Contract No. NaW-64-0178-s), Urbana, IL (1965)
10. Kenane, M., Benzeggagh, M.L.: Measurement of mixed-mode delamination fracture toughness of unidirectional glass/epoxy composites with mixed-mode bending apparatus. Compos. Sci. Technol. **56**(4), 439 (1996)
11. Reeder, J.R.: 3D mixed mode delamination fracture criteria: an experimentalist's perspective. In: Proceedings of American Society for Composites, 21st Annual Technical Conference, Dearborn, MI, 2006
12. Fenner, J.S., Daniel, I.M.: Fracture toughness and impact damage resistance of nanoreinforced carbon/epoxy composites. In: Proceedings of SEM Annual Conference, Costa Mesa, CA, 8–11 June 2015
13. Schwab, R.C.: Use of tapered double cantilever beam specimens for fatigue crack growth studies. Mech. Eng. **91**(8), 71 (1969)
14. Fenner, J.S., Daniel, I.M.: Hybrid nanoreinforced carbon/epoxy composites for enhanced damage tolerance and fatigue life. Compos. Part A **65**, 47–56 (2014)

# Chapter 11
# Characterization of Fatigue Induced Damage Evolution in CFRPs Using DIC

**H. Murthy and S. Venkatachalam**

**Abstract** Damage evolution in fatigue tests (R = 0.5, −1, 2) conducted on carbon fiber reinforced plastic (CFRP) composites has been characterized using digital image correlation (DIC). Since damage initiation/delamination is a local phenomenon affecting transverse strain more than the longitudinal, local transverse strain is a better indicator of onset of delamination and its propagation. Variation of transverse strain near the initiated delamination with cycles indicates that the damage evolution occurs over 2–3 stages. Each stage has a stable damage growth with sudden increase between the stages. Waviness and the associated error due to the lag between image and load data acquisition was overcome by plotting the maximum transverse strain obtained from a curve fit to each set of continuous cycles. Error due to large relative deformations was avoided by choosing different reference images for different stages. Extent of damage zone and its evolution was characterized by the length over which the transverse strain exceeds a limiting value, which was taken to be that at the end of first stage in the plot. Rate at which the damage propagates shows similar variation as that of the local transverse strain, which shows that the latter can be used as an indicator of fatigue damage evolution. This also provides a method to quantify the damage in terms of local transverse strain, which can in turn be used to validate any developed damage models.

**Keywords** CFRP • Fatigue • Damage evolution • DIC • Transverse strain

## Nomenclature

| | |
|---|---|
| $\sigma_x$ | Longitudinal stress |
| $\epsilon_y$ | Transverse strain |
| $\sigma_{max}$ | Maximum stress |
| $\Delta\epsilon$ | Longitudinal strain range |
| $\Delta\sigma$ | Longitudinal stress range |

## 11.1 Introduction

The applications of fiber reinforced composite materials in engineering industries have been steadily increasing due to their high strength to weight ratio and stiffness to weight ratio. Since components made of composite materials are subjected to different types of loading, it is important to study the mechanical behaviour of composite materials especially under cyclic loading. Unlike in metals, the damage behaviour in composite materials is complex due to high modulus ratio between the fiber and matrix, volume fraction of the individual constituents, fiber orientation, lay-up sequence and fabrication process [1]. In composite materials, the damage occurs due to various mechanisms like cracks developed in matrix, de-bonding between fiber and matrix, delamination, fiber fracture/buckling [2, 3] etc. The damage occurs in different stages. During each stage, one or a few of the mentioned damaged modes may dominate. Further, these damage modes get influenced by the choice of matrix and fiber, the adhesion characteristics between them and the manufacturing process. Therefore, each composite may have a different mechanism of damage evolution and eventual failure. Due to these reasons, phenomenon of fatigue damage in composites is very poorly understood and the models available for prediction of failure lives are not yet

H. Murthy (✉) • S. Venkatachalam
Department of Aerospace Engineering, Indian Institute of Technology Madras, Chennai, India
e-mail: hsnmurthy@gmail.com



A.T. Zehnder et al. (eds.), *Fracture, Fatigue, Failure and Damage Evolution, Volume 8*,
Conference Proceedings of the Society for Experimental Mechanics Series, DOI 10.1007/978-3-319-42195-7_11

robust. Consequently, relatively higher factor of safety is used in industry due to lower confidence in the predicted life of a composite structure subjected to cyclic load.

Typically, the fatigue damage in composites was assessed by reduction in global stiffness in the longitudinal direction or reduction in Poisson's ratio. The reduction in global stiffness in the longitudinal direction was not significant to characterize the fatigue behaviour [4]. Smith and Wood [5] conducted static tensile experiments on symmetric cross ply GFRP and CFRP and used 0/90 rosette to find the longitudinal and transverse strains. They reported that the reduction in Poisson's ratio is much more than the reduction in longitudinal modulus, which is less than 5 % in CFRP. Surgeon et al. [6] also conducted tensile experiments on variety of cross ply CFRP laminates which are attached with 0/90 strain gauge and found that a little or no reduction in Young's modulus but large reduction in Poisson's ratio. Paepegem et al. [7] conducted fatigue experiments on cross ply E-glass fiber/epoxy laminates and observed that the degradation of Poisson's ratio is much greater than that of the longitudinal stiffness. They used an extenso-meter to measure the longitudinal strain and a strain gauge to measure the transverse strain. Hence, in the present study, reduction in local transverse stiffness of composite material is considered as a parameter to estimate the damage induced by fatigue. But the use of strain gauge to find local strain in transverse direction has the disadvantage of providing strain at a location where the strain gauge is affixed and it is difficult to find the strain at the damage zone as it is not known a priori. Moreover, it is not possible to apply load beyond the elastic limit of the strain gauge material. Consequently, it is essential to have a damage monitoring technique, which provides whole field information and is sufficiently quantitative. Digital Image Correlation (DIC) is one such non-contact technique, which provides the whole field strain data on a surface. Since the damage in composites have been found to always initiate on the free surface due to lesser constraints, DIC can be used to obtain the strains on this free surface and hence monitor the damage evolution due to cyclic loading.

## 11.2 Fatigue Testing of CFRP

In this work, elaborate well-controlled experiments have been conducted to understand fatigue damage in CFRP composites. Fatigue tests have been conducted at three different stress ratios (R = 0.5, 2 and −1). The specimens used for fatigue test are 17 layered carbon fiber (HS Carbon UD Fabric G0827-B1040-HP03-1 F) and epoxy resin (Epolam 2063) composites with the stacking sequence of $[+45/90/-45/0/+45/0/-45/0/90]_S$. The gripping sections of the specimens are strengthened by adding glass fiber reinforced plastic tabs to avoid failure at the gripping portion. The dimensional details of the specimen used for tension-tension, compression-compression and tension-compression fatigue tests are shown in Fig. 11.1. Specimens used for compression mode fatigue tests have shorter gage length to avoid buckling. The specimens were supplied by National Aerospace Laboratory, Bangalore.

Fatigue tests have been conducted using a *100 kN INSTRON 8801* fatigue testing machine. The experimental setup (photograph shown in Fig. 11.1) consists of two cameras, one on either side of the specimen, to acquire images of the speckle pattern introduced on the specimen and four light sources (LED lamps) to illuminate the specimen surface. For fatigue tests involving compressive loads (C-C and T-C), two support plates with teflon pads were mounted, one on either side of the actuator, to avoid bending of the actuator and to ensure that the specimen is loaded axially. Silicone grease was applied to

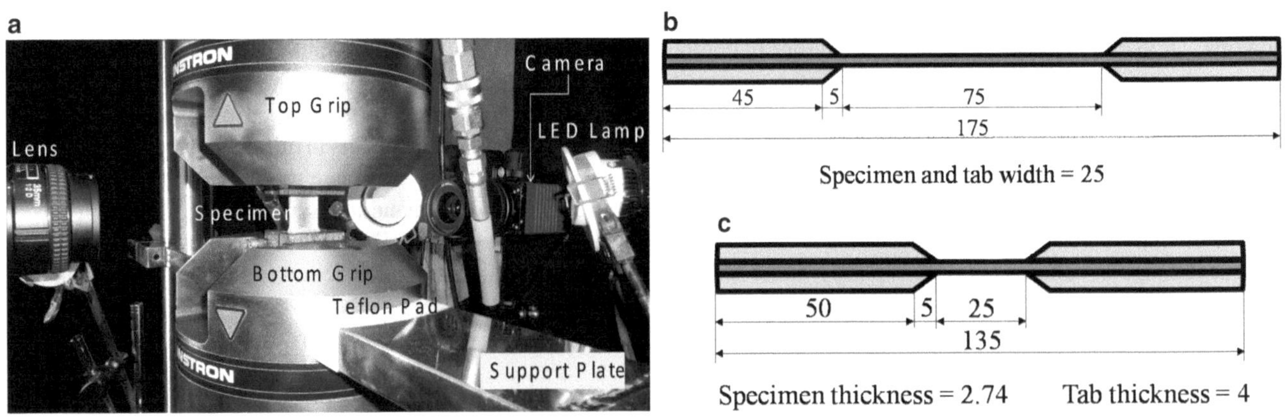


**Fig. 11.1** Fatigue testing set-up with DIC for full field strain measurements and specimens used for fatigue tests at different R-ratios. All dimensions are in mm. Specimen and tab width = 25 mm. Specimen thickness = 2.74 mm. Tab thickness = 4 mm

reduce the friction between the teflon pads on the support plates and the bottom grip. The specimen is gripped between hydraulic wedge grips of size 50 × 50 mm with flat carbide coated surf alloy jaw faces.

Speckles are introduced on the specimen through spray painting and these speckle patterns are tracked with time to obtain strains on the surface using DIC. It is a non-contact full field image analysis technique for measuring displacement and strain from relative motion of the speckle patterns. This technique correlates the gray values on digital images captured before and after the deformation of the specimen to obtain the full field strain. For unique correlation, the technique compares the gray intensity over a small square region of pixels called subset instead of a pixel. By tracking the subset on the digital images, the system measures the average surface displacement and strain of a pixel at subset centre. In similar fashion, all the subsets of the digital images can be tracked to find the full field surface displacement and strain [8].

## 11.3 Damage Monitoring Through Transverse Strain Measurement

During preliminary tension-tension fatigue tests, it was found that there is less than 2 % reduction in the global longitudinal stiffness even after considerable visible delamination in the specimen. This may be because damage initiates locally and may not modify global longitudinal stiffness significantly until it is critical. Moreover, it is observed that plies do bend locally in transverse direction after delamination. Hence, it has been assumed that change in local transverse stiffness along the thickness direction is a better indicator of delamination. To evaluate the local transverse stiffness, it is required to measure the local transverse strain in the specimen. Since the location of damage initiation is not known a priori, DIC technique has been used to get the full field strain data. This strain data would lead to the identification of damage zone on the free surface along the thickness direction, over the gage length of the specimen.

### *11.3.1 Digital Imaging and Damage Zone Detection*

For strain measurement using DIC, images of speckled lateral faces along the thickness direction of the composite specimen have been captured at specified frame rate using two charge-coupled device (CCD) cameras during fatigue tests. For fatigue experiments with frequency, f = 1 Hz, 10 images of each lateral surface are captured per cycle with a dwell period of 99 cycles. For fatigue experiments with f = 5 Hz, 6 images of the each lateral surface are captured per cycle for two cycles with a dwell period of 198 cycles. These images of speckled patterns are then correlated with image of the speckle pattern on the specimen in undeformed condition. From this correlation, transverse strain contour plots are obtained. Any significant local damage in the specimen would reflect in these contour plots. The damage zone/area on the lateral faces along the thickness direction of the composite specimen were identified by observing these transverse strain contour plots and then the transverse strain at the location was tracked as a function of number of cycles to understand the damage evolution with number of cycles.

### *11.3.2 Evolution of Maximum Local Transverse Strain Measured by DIC*

Figure 11.2a shows the transverse strain contour plot for a typical tension-tension experiment (R = 0.5, $\sigma_{max}$ = 600 MPa and f = 1 Hz). After identifying the damage zone from this plot, a point (i.e., a pixel) close to the damage zone is considered as a critical point for finding strain variation with cycles. The local transverse strain of this critical point has been extracted from the first image of each cycle (10 images captured/cycle) for the entire test. The variation of this local transverse strain divided by the longitudinal stress is plotted against the number of cycles (Fig. 11.2b). Small waviness and large scatter beyond 40,000 cycles can be observed in the shown plot. The large scatter is observed only after visible delamination, which in turn leads to large local transverse strain [9]. The waviness may be due to the time lag between the load data acquisition and image captured. To avoid this lag, peak strain was estimated by fitting a sinusoidal curve to the strains obtained from DIC of images and peak load was obtained from the data acquired during each cycle. Ratios of these peaks were used for plotting instead of loads and strains at any random instant of time. For obtaining the best fit sine curve, the strain variation for each cycle was assumed as follows:

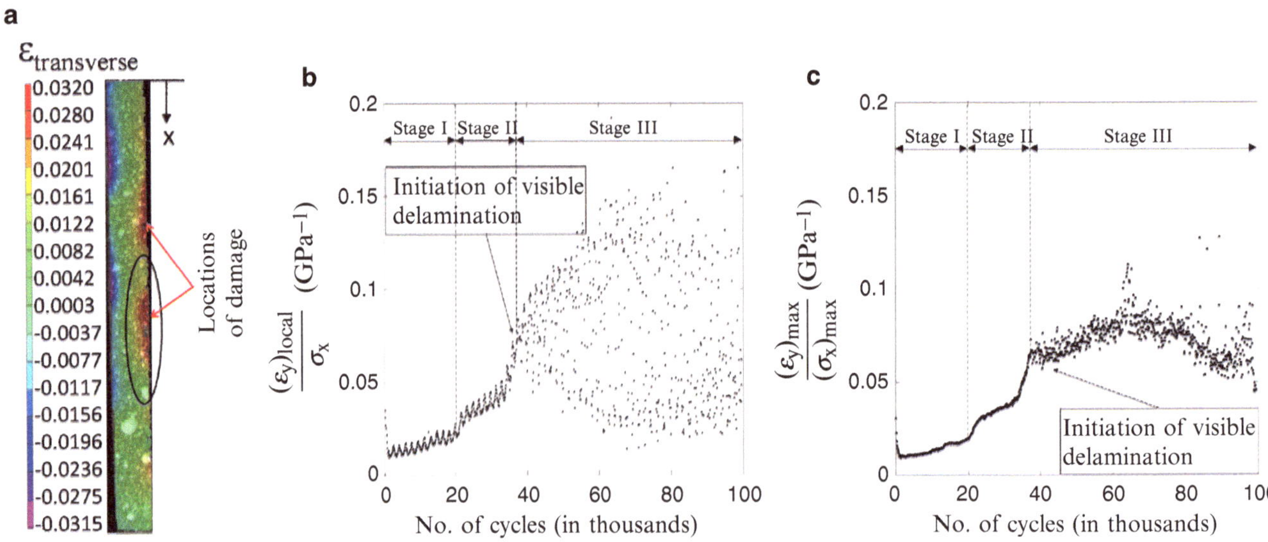


**Fig. 11.2** (**a**) Transverse strain contours of the specimen at 37,000 cycles; Variation of (**b**) $(\epsilon_y)_{local}/\sigma_x$; (**c**) $(\epsilon_y)_{max}/(\sigma_x)_{max}$ with number of cycles for the fatigue test with R = 0.5, $\sigma_{max}$ = 600 MPa and f = 1 Hz with undeformed state as reference image

$$Y = A + \lambda \sin(\omega t + \phi) = A + B\sin(\omega t) + C\cos(\omega t) \tag{11.1}$$

where A is mean strain, λ is strain amplitude, ω is the frequency of strain data acquired and ϕ is phase angle. Linearization of the equation with respect to A, B and C as shown gives B = λcos(ϕ) and C = λsin(ϕ). The values of A, B and C can be obtained from linear fit using least square method for different ω. Value of ω could be obtained by minimizing the error using these fitted parameters for different chosen values of ω.

The variation of ratio of maximum local transverse strain to maximum applied longitudinal stress is as shown in Fig. 11.2c. It can be observed that the waviness has been eliminated by plotting the maximum local transverse strain variation with cycles. The change of slope in strain variation is indicated as different stages in the plot. These stages may correspond to different damage modes/mechanisms in the composites during fatigue. Further, the cycle at which the visible delamination, as observed from the images, occurs was found to correlate with the onset of scatter in the plot. It can be concluded that the variation of local maximum transverse strain in the specimen during fatigue cycling clearly indicates the loss of stiffness even before the visible delamination occurs.

Though the waviness has been eliminated in the plot, still there is scatter beyond 40,000 cycles. This could be due to large relative displacement between the speckle patterns in the underformed configuration and the deformed configuration. To remove the scatter in the plot, the images have been correlated again using DIC software with different reference images to obtain the relative strains. In this analysis, three stages have been considered. Image of the undeformed specimen was used as reference image for stage I (0–20,000 cycles). For other two stages (20,000–37,000 cycles and 37,000 cycles to end of the experiment), last image of the previous stage was used as the reference image. The absolute strain values at each stage have been obtained by adding the relative strains of that stage to the absolute strain value of the reference image already obtained from the analysis of images in the previous stage. From these absolute strains, maximum transverse strains in each cycle have been obtained using sine curve fit. Figure 11.3 shows the variation of ratio of maximum local transverse strain to maximum applied longitudinal stress with number of cycles. It can be observed that this strain plot has smooth variation and the scatter is eliminated as well.

## 11.4 Experimental Results and Discussion

To confirm that the local transverse compliance taken as a ratio of local transverse strain at a critical point close to the damage to the applied stress can be a good indicator of fatigue damage growth in composites, it was decided to measure the delamination growth near the damage zone with increase in number of cycles.

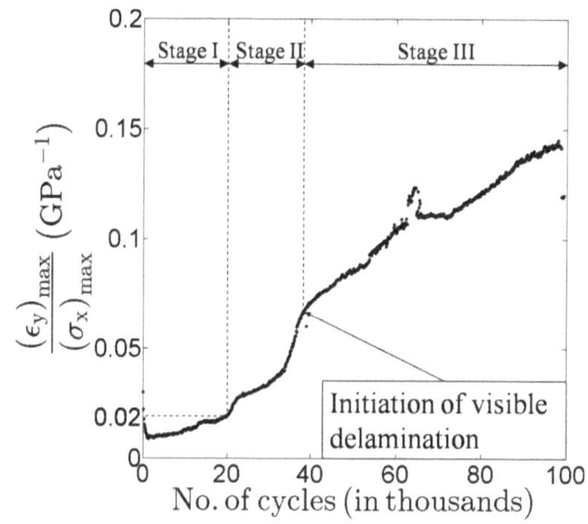


**Fig. 11.3** Variation $(\epsilon_y)_{max}/(\sigma_x)_{max}$ with number of cycles for a fatigue test with $R = 0.5$, $\sigma_{max} = 600$ MPa and $f = 1$ Hz

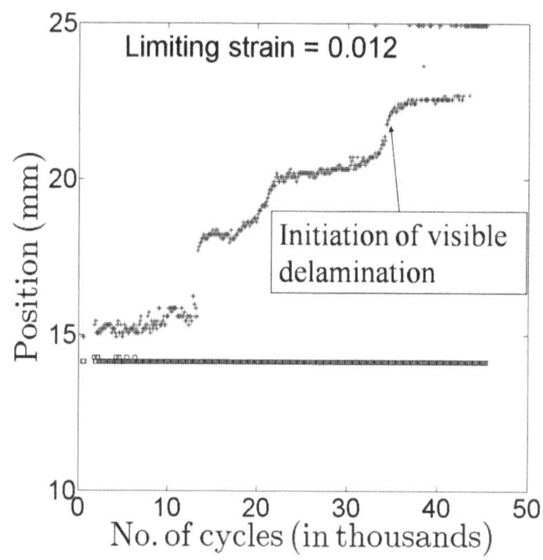


**Fig. 11.4** Increase in damage length with number of cycles for a fatigue test with $R = 0.5$, $\sigma_{max} = 600$ MPa and $f = 1$ Hz

### *11.4.1 Tension-Tension Fatigue Test*

For the tension-tension fatigue test discussed in the above section, the delamination growth near the damage zone with increase in number of cycles was measured as shown in Fig. 11.4. To find the damage growth, the length over which the strain exceeds a limiting value has been considered with increase in number of cycles. This limiting value of strain was obtained from the end of stage I (0.012) as shown in Fig. 11.3. It is seen that the delamination growth also shows a trend similar to that of the local transverse compliance (Fig. 11.3) with three distinct stages of damage growth. This provides confidence that the local transverse compliance taken as a ratio of local transverse strain at a critical point close to the damage to the applied stress is indeed a good indicator of fatigue damage growth in composites. Further, this local transverse compliance provides a quantity that can be compared with while verifying the models developed to predict fatigue damage evolution.

To confirm the method of monitoring the damage by measuring the local maximum transverse strain, further fatigue tests have been conducted with other R-ratios ($R = 2$ for compression-compression and $R = -1$ for tension-compression fatigue tests). Results of all these fatigue tests are discussed below.

### *11.4.2 Compression-Compression Fatigue Test*

The variation of ratio of maximum transverse strain to maximum applied longitudinal stress versus the number of cycles for the fatigue test at $R = 2$, $\sigma_{max} = 424$ MPa and $f = 5$ Hz is shown in Fig. 11.5. The initiation of visible delamination and catastrophic failure of the specimen occurs at about 67,000 cycles. From these experiments, it was also observed that there is

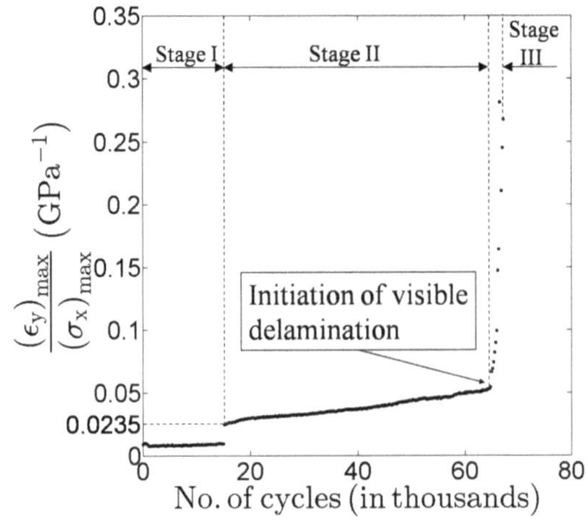


**Fig. 11.5** Variation $(\epsilon_y)_{max}/(\sigma_x)_{max}$ with number of cycles for a fatigue test with R = 2, $\sigma_{max}$ = 424 MPa and f = 5 Hz

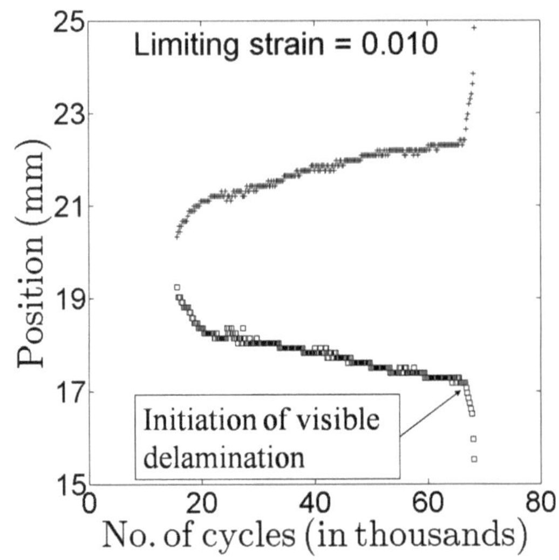


**Fig. 11.6** Increase in damage length with number of cycles for a fatigue test with R = 2, $\sigma_{max}$ = 424 MPa and f = 5 Hz

no significant difference between the cycles corresponding to initiation of visible delamination and final failure of the material unlike the tension-tension fatigue experiments, where the specimen was capable of carrying the load for very high number of cycles even after the initiation of visible delamination. This may be mainly due to the fact that the composites are fiber controlled and hence have low load carrying capability in compression mode after the initiation of visible delamination. Figure 11.6 shows increase in damage length with increase in number of cycles for compression-compression fatigue tests. For this case, the strain of 0.01 (value at the end of stage I) was taken as the limiting value. In this case also, delamination growth shows a trend similar to that of the local transverse compliance.

### *11.4.3 Tension-Compression Fatigue Test*

Figure 11.7 shows the variation of ratio of maximum transverse strain to maximum applied longitudinal stress versus the number of cycles for the fatigue test conducted at R = −1, $\sigma_{max}$ = 214 MPa and f = 5 Hz. The specimen failed after 700,000 cycles of loading. The variation of these parameters clearly indicates the three stages of damage evolution during fatigue cycling. Figure 11.8 shows increase in damage length with increase in number of cycles for tension-compression fatigue tests. Behaviour is similar to tension-tension (R = 0.5) fatigue tests, though the third stage is much shorter. Again, this may be because, once there is a significant damage, it might grow fast due to bending effect (local buckling) under compressive loading.

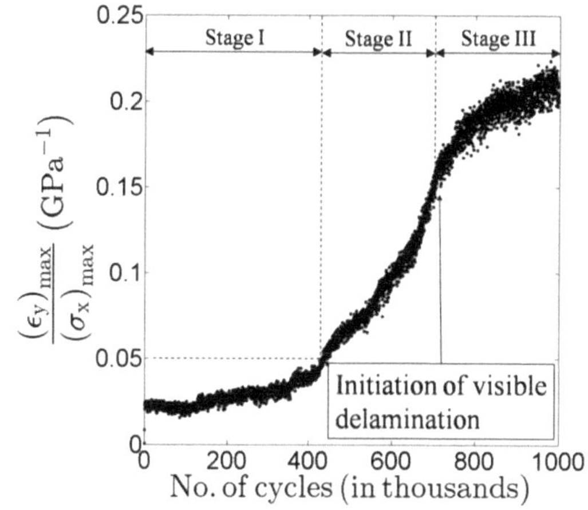


**Fig. 11.7** Variation $(\epsilon_y)_{max}/(\sigma_x)_{max}$ with number of cycles for a fatigue test with $R = -1$, $\sigma_{max} = 214$ MPa and $f = 5$ Hz

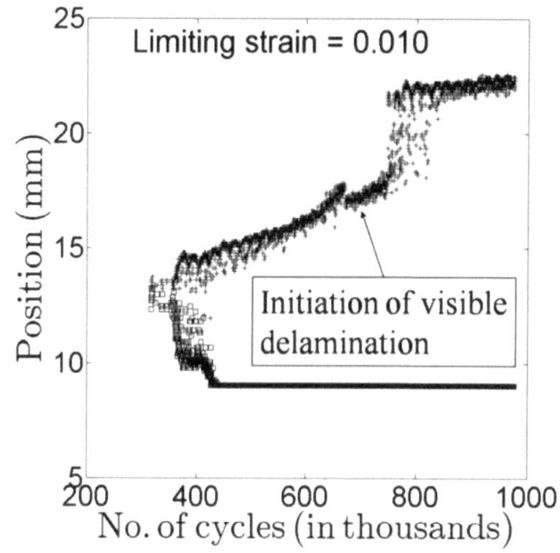


**Fig. 11.8** Increase in damage length with number of cycles for a fatigue test with $R = -1$, $\sigma_{max} = 214$ MPa and $f = 5$ Hz

## 11.5 Conclusions

Damage evolution in CFRP materials subjected to fatigue loading has been studied using DIC. The variation of local transverse strain with number of cycles was found to be a better indicator of loss in stiffness in the CFRP material due to delamination and related damage mechanisms. The waviness and scatter in the plot of ratio of transverse strain to applied stress with number of cycles were avoided using sine curve fit to obtain maximum transverse strain in a cycle and different reference images for DIC respectively. Variation of the local transverse stiffness with number of fatigue cycles indicates that the damage evolution occurs over 2–3 stages. Further, the local transverse compliance taken as a ratio of local transverse strain at a critical point close to the damage to the applied stress can be a good indicator of fatigue damage growth in composites is confirmed by measuring the delamination growth near the damage zone with increase in number of cycles.

**Acknowledgments** The authors gratefully acknowledge National Aerospace Laboratories, Bangalore for the fabrication of the tested specimens.

## References

1. Bhat, M.R., Murthy, C.R.L.: Fatigue damage stages in unidirectional glass-fibre-epoxy composites: identification through acoustic emission technique. Int. J. Fatigue **5**, 401–405 (1993)
2. Jones, R.M.: Mechanics of Composite Materials, 2nd edn. Taylor & Francis, Philadelphia (1999)
3. Reifsnider, K.L., Case, S.W.: Damage Tolerance and Durability of Material Systems. Wiley, New York (2002)
4. Bourchak, M., Farrow, I.R., Bond, I.P., Rowland, C.W., Menan, F.: Acoustic emission energy as a fatigue damage parameter for CFRP composites. Int. J. Fatigue **29**, 457–470 (2007)

5. Smith, P.A., Wood, J.R.: Poisson's ratio as a damage parameter in the static tensile loading of simple crossply laminates. Compos. Sci. Technol. **38**, 85–93 (1990)
6. Surgeon, M., Vanswijgenhoven, E., Wevers, M., Van Der Biest, O.: Transverse cracking and Poisson's ratio reduction in cross-ply carbon fibre-reinforced polymers. J. Mater. Sci. **34**, 5513–5517 (1999)
7. Van Paepegem, W., De Baere, I., Lamkanfi, E., Degrieck, J.: Poisson's ratio as a sensitive indicator of (fatigue) damage in fibre-reinforced plastic. Fatigue Fract. Eng. Mater. Struct. **30**, 269–276 (2007)
8. Sutton, M.A., Orteu, J.-J., Schreier, H.W.: Image Correlation for Shape, Motion and Deformation Measurements. Springer, New York (2009)
9. Khaja Mohiddin, S.M., Subramanian, S.J., Murthy, H.: Study of damage evolution during fatigue of composites using DIC. Proceedings of SEM XII International Congress and Exposition on Experimental and Applied Mechanics Conference, Costa Mesa, USA, CD-ROM (2012)

# Chapter 12
# Damage Characterization for Electronic Components Under Impact Loading

**Sangwook Sihn, Christie L.H. Devlin, Steven R. Dooley, Ajit K. Roy, and Eric R. Heller**

**Abstract** A new experimental method has been develop to characterize critical interfacial damage parameters of solder interconnects subjected to high strain-rate mechanical loading, simulating shock or impact loading. A test apparatus and a test specimen were devised to experimentally characterize such critical damage parameters, particularly interfacial shear strength and fracture toughness. The test fixture was designed to easily mount and unmount test specimens, and accommodate various sizes of electronic components and solder layer thicknesses. Test specimens were fabricated with both metallic and polymeric solder materials, and tests were conducted under various shear load rates. It was found that both strength and fracture toughness exhibit significant rate-dependency.

**Keywords** Solder failure • Impact loading • Interfacial damage • Finite element analysis • Shear failure

## 12.1 Background and Introduction

Damage tolerance of electronic components subject to severe impact loading is of a great concern for reliable electronic operations. Under severe impact loading, the device can fail mechanically and electrically due to loss of adequate contact and bonding between the device package and the printed circuit board (PCB). One of standard methodologies for understanding structural responses and damage mechanisms of the electronic packages under the impact loading is the board-level drop test method proposed by Joint Electron Device Engineering Council (JEDEC) [1–3]. Numerous studies have conducted the board-level drop tests with various kinds of electronic components [4–6]. Among them, Kaplan et al. conducted experimental and computational investigation of dynamic response of PCB strain under the high-g shock impact loading [7]. In this study, a strain gauge was mounted on the PCB to monitor the dynamic strain response under ~15,700 g vertical acceleration, resulting in a maximum strain amplitude peak of near 6,300 μm. It was noted that the strain response of the PCB was sinusoidal with decaying amplitudes over time, which indicated that the major mode of the PCB deformation was longitudinal bending perpendicular to the plane of circuit board. Such strain responses were also reported by other experimental investigations [5, 8]. Due to the bending stiffness difference between the PCB and the electronic packages on it, tensile and shear stresses are built up at their interfaces during the PCB bending, causing interfacial cracks to open and slide, resulting in component failure. Therefore, solder joint reliability is closely related to the dynamic responses of PCB, and further the interfacial damage resistance between the PCB and the electronic components. Numerous experimental and computational studies regarding the impact damage indicate that interfacial normal and shear fractures near the PCB-package interfaces are the dominant damage mechanisms [6, 9–15]. Therefore, it is essential to characterize the interfacial damage resistance parameters to assess and predict the overall performance of such electronic components under impact loading.

S. Sihn (✉)
Materials and Manufacturing Directorate, Air Force Research Laboratory, Wright-Patterson Air Force Base, Dayton, OH 45433, USA

Nonstructural Materials Division, University of Dayton Research Institute, 300 College Park, Dayton, OH 45469-0050, USA
e-mail: sangwook@alumni.stanford.edu

C.L.H. Devlin
Sensors Directorate, Air Force Research Laboratory, Wright-Patterson Air Force Base, Dayton, OH 45433, USA

Wyle Laboratories, Dayton, OH, USA

S.R. Dooley
Sensors Directorate, Air Force Research Laboratory, Wright-Patterson Air Force Base, Dayton, OH 45433, USA

A.K. Roy • E.R. Heller
Materials and Manufacturing Directorate, Air Force Research Laboratory, Wright-Patterson Air Force Base, Dayton, OH 45433, USA



A.T. Zehnder et al. (eds.), *Fracture, Fatigue, Failure and Damage Evolution, Volume 8*,
Conference Proceedings of the Society for Experimental Mechanics Series, DOI 10.1007/978-3-319-42195-7_12

In the present study, we devised a new test apparatus and a test specimen to experimentally characterize the critical interfacial damage parameters of solder interconnects. The test apparatus was devised to experimentally characterize such critical damage parameters, particularly interfacial shear strength and fracture toughness. A measure of fracture toughness in terms of strain energy release rate was evaluated by means of a simplified shear test apparatus. The test apparatus was designed to easily mount and unmount test specimens, and accommodate various sizes of components and solder layer thicknesses. Test specimens were fabricated with both metallic and polymeric solder materials, and tests were conducted under various shear load rates. Rate dependency of the fracture toughness, strength and modes of the failed solder joints were also examined. Once measured, these parameters can be used for the computational simulations to predict the damage phenomena under high-g impact loading. We also developed FE models based on the cohesive zone model (CZM) to assess the feasibility of the present method.

## 12.2 Experimental Setup

The test specimen was fabricated with the PCB and an electronic component (die/chip) bonded to the top surface of the PCB with solder materials. The PCB is fabricated with FR-4 (woven glass-fiber epoxy composites). For feasibility study of the proposed test method and apparatus, a dummy chip was made of the same material as the PCB. Fig. 12.1 shows schematic top and side views of the test specimen and its dimensions. The PCB has two holes to be screwed to the back support grip of the test apparatus, as Fig. 12.2c shows. Length ($L_B$) and width ($W_B$) of the PCB are 25.4 mm (1.0 in.) by 12.7 mm (0.5 in.). Length ($L_C$) and width ($W_C$) of the dummy chip are 3.81 mm (0.15 in.) by 3.81 mm (0.15 in.). The thicknesses of the PCB ($T_B$) and the dummy chip ($T_C$) are approximately 1.6 mm (63 mil). Thin Cu pads, whose size dimensions are 2.54 mm (0.1 in.) by 2.54 mm (0.1 in.), and thickness is approximately 25 μm, are adhered to the PCB and the chip surfaces for better solder bonding. Exact dimensions of $L_B$, $W_B$, $T_B$, $L_C$, $W_C$, of each specimen were measured using a caliper before the tests. Furthermore, a total thickness ($T_T$) was measured from the top surface of the chip to the bottom surface of the PCB, as Fig. 12.1b shows. The thickness of the solder layer ($T_S$) was then calculated as $T_S \cong T_T - 2T_B$, where $T_C \cong T_B$.

The shear test apparatus was fabricated based on a design modeled with several iterations of numerical simulations using a commercial finite element (FE) software program ABAQUS. Fig. 12.2 shows schematic views of the shear test apparatus, which consists of a back-support grip (aluminum), a shear-tool grip (aluminum), and a pin plug (stainless steel). Holes in cylindrical portion of the back-support grip and the shear-tool grip are pinned to test load frame (MTS) grips, as shown in Fig. 12.3. While the back-support grip is fixed, the shear tool is horizontally displaced to apply the shear load to the die/PCB interface, and thus to induce pure interfacial shear failure. To prevent undesirable bending of PCB during the shear loading, the PCB is mounted and screwed to a flat surface of the back-support grip with two screws, as shown in Fig. 12.2c. The edge of tested die/chip soldered on the PCB surface is in contact with the pin plug, as shown in Fig. 12.2c. The pin plug is inserted through a hole in the shear-tool grip, as shown in Fig. 12.2a, b. With this design, we can easily reutilize the back-support and shear-tool grips for multiple tests, and furthermore can have flexibility of testing various sizes of dies/chips as well as solder layers.

The shear test apparatus and specimen were modeled with ABAQUS to conduct the shear loading simulations and to study the correlation between the load-displacement behaviors and the interfacial fracture toughness. The actual material properties and dimensions of the apparatus and specimens were used in the model. The cohesive zone model (CZM) based

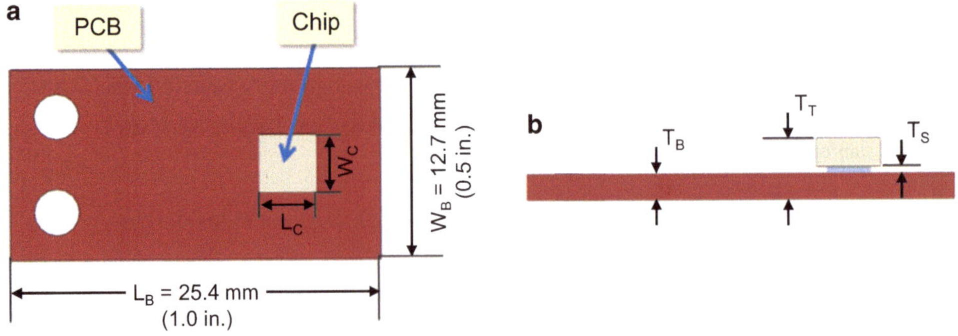


**Fig. 12.1** Schematic (**a**) *top* and (**b**) *side* views of a shear test specimen and its dimensions with a dummy chip soldered to a PCB with Cu pads in-between

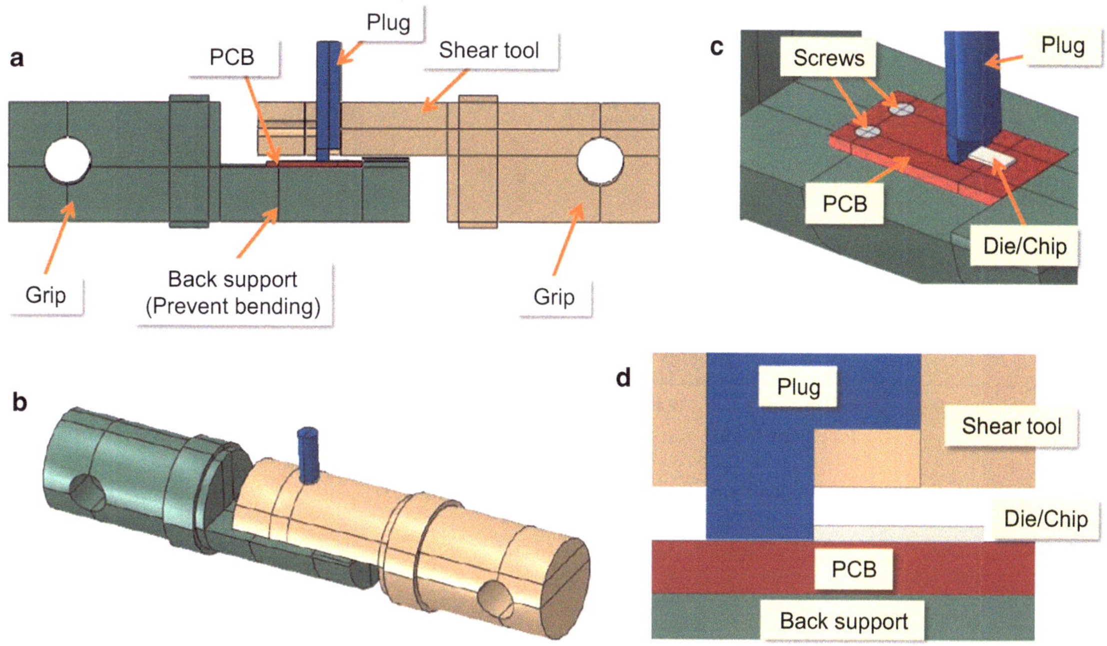


**Fig. 12.2** Schematic views of shear test apparatus and test specimen

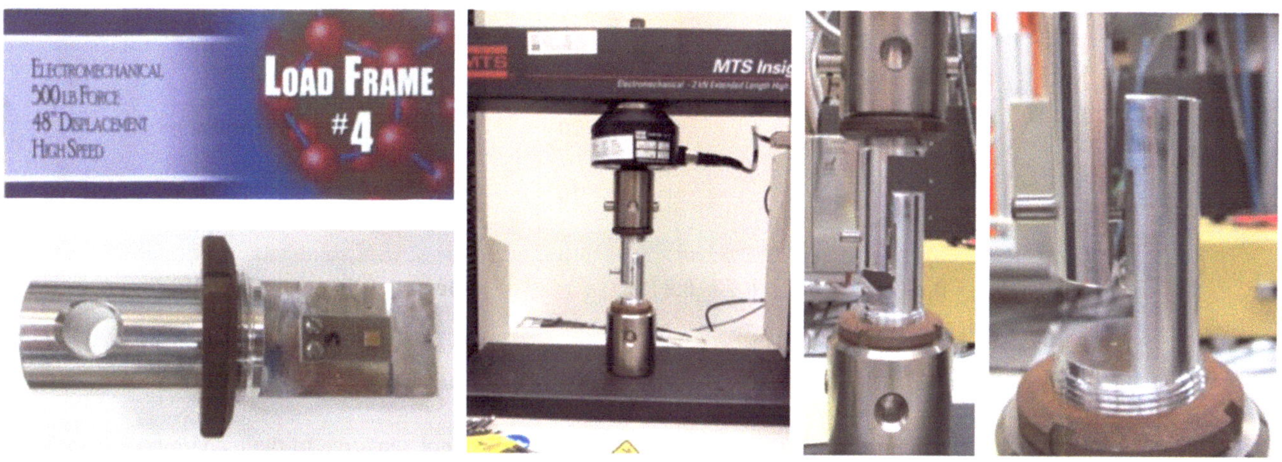


**Fig. 12.3** Shear test apparatus and specimen mounted and tested in an MTS load frame

on the traction-separation law was implemented to predict the initiation and progression of damage at the interfaces near the solder layer. Several model parameters required by the CZM simulation are not readily available or difficult to measure, so reasonably assumed values were used for these parameters, and parametric studies were conducted to determine the sensitivity of these parameters on the final results. The assumed CZM parameters and their reference values are as follows:

- Normal and shear stiffnesses: $k_n = k_s = 160$ MPa
- Peak nominal stress (normal): $t_n = 20$ MPa
- Peak nominal stress (shear): $t_s = 30$ MPa
- Normal mode fracture energy: $G_n = 2000$ N/m
- Shear mode fracture energy: $G_s = 6000$ N/m
- Friction coefficient (tangential): $f = 0.3$
- Solder layer thickness: t = 0.368 mm (Obtained from caliper measurements)
- Damage evolution: Exponential softening law.

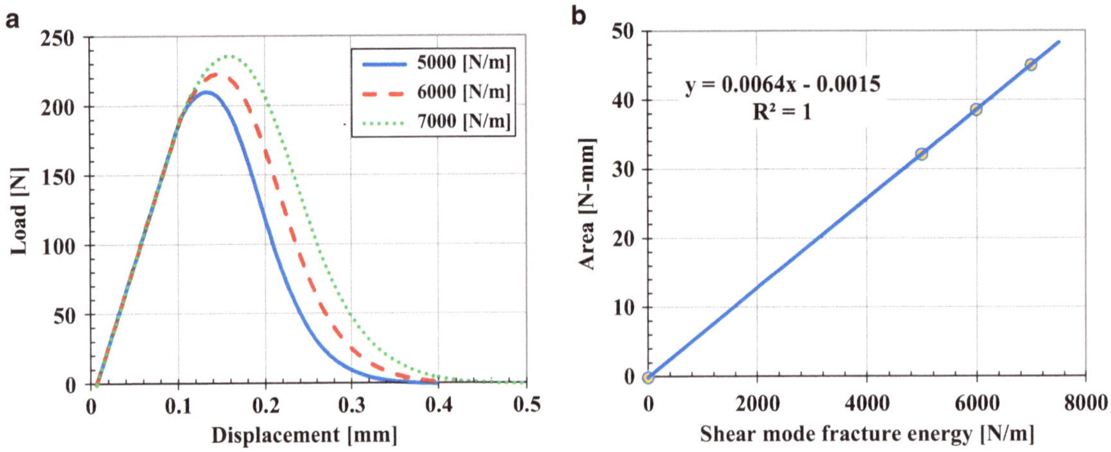


**Fig. 12.4** (**a**) Load-displacement curves simulated with three shear mode fracture energy ($G_s = 5000, 6000, 7000$ N/m); (**b**) Area underneath the load-displacement curves against $G_s$

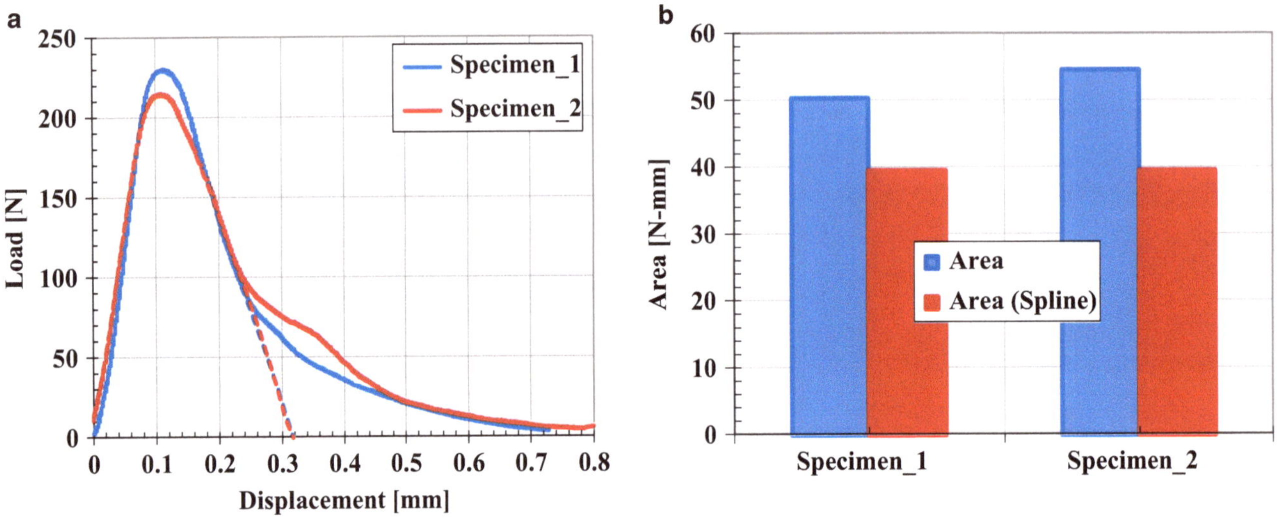


**Fig. 12.5** (**a**) Load-displacement curves; (**b**) Area underneath the curves of two shear test specimens

It was found from several parametric studies that the simulation results are not sensitive to the CZM model parameters except the shear mode fracture energy ($G_s$). Therefore, we can conclude that the present test apparatus and specimens are suitable to characterize the interfacial shear mode fracture toughness.

Three values (5000, 6000 and 7000 N/m) of the interfacial shear mode fracture parameter, $G_s$, were chosen for the parametric study, and the resulting load-displacement curves are plotted in Fig. 12.4a, b shows the calculated area underneath the load-displacement curves against the three $G_s$. The three points as well as the origin (0, 0) are fit with a linear interpolation, whose fitting equations and the reliability of the fit are shown in the figures. All three cases indicate that the area underneath the load-displacement curves are linearly correlated with $G_s$. Therefore, once the area is obtained from experimentally measured load-displacement curves using the proposed shear test apparatus and specimens, $G_s$ value can immediately be extracted from the fit equation.

## 12.3 Results and Discussion

Tests were conducted in the MTS machine at various displacement rates. During the tests, load and displacement were recorded. Fig. 12.5a shows an example of the load-displacement curves (solid lines) of the two shear specimens tested at 4.23 μm/s (0.01 in/min). The specimens were fabricated with metallic tin-lead (Sn63/Pb37 from Chip Quik) solder. Both

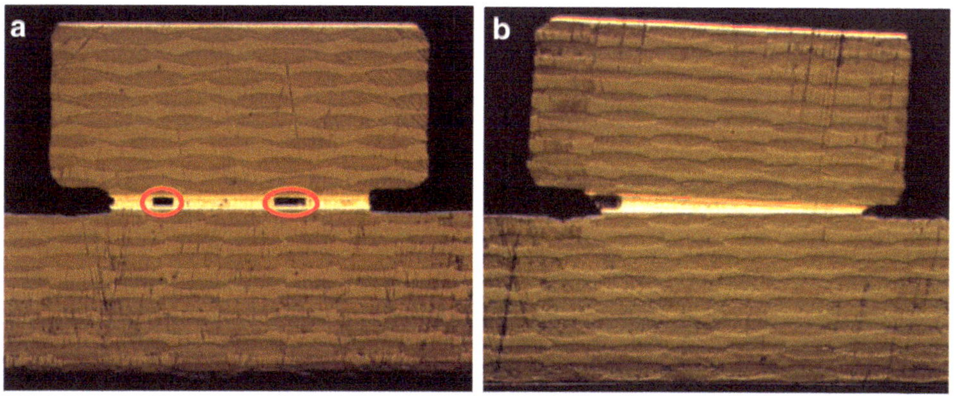


**Fig. 12.6** Optical microscopic cross section images of sectioned shear test specimens showing (**a**) voids in the solder layer (highlighted with *red circles*); (**b**) significantly tilted chip

**Fig. 12.7** Microscopic images of failed sections of (**a**) cohesive failure at the solder layer; (**b**) interfacial failure between PCB and Cu pad on PCB surface; (**c**) mixed cohesive and interfacial failure

curves show that the load increases with the increment of the shear deformation until a maximum load, where the interfacial damage likely initiates at the solder layer. The load decreases monotonically with the further increase of the displacement. Note that both curves are nearly identical in the incremental portion up to approximately 200 N and 0.07 mm with a slight difference in the maximum load. The curves also show the similar decremental behavior up to approximately 100 N at 0.24 mm. However, with the further increase of the displacement (>0.24 mm), the curves become significantly different. The curve of Specimen_1 continues to decreases monotonically, while the curve of Specimen_2 has a kink at the displacement between 0.24 and 0.46 mm. Both curves finally decrease nearly identically for larger displacements than 0.46 mm.

The difference in the load-displacement behavior, especially in the load-decremental portion, might be attributed from material and fabrication variability, such as void defect in the solder layer, misalignment/tilt of the chip, etc. To study such variability, the specimens were sectioned and polished to inspect their cross sections near the solder layer using an optical microscope. Optical microscopy images of two sectioned specimens in Fig. 12.6 clearly show the void defect in the solder layer and misalignment/tilt of the chip, which attribute to the variation in the load-displacement curves and damage behavior. To isolate the consequences of the material and fabrication variability from the overall trends, the decremental portion of the curve was simplified by ignoring the portion beyond any major kinks or inflections, and fitting the rest of the curves with a spline function. Dotted lines in Fig. 12.5a indicate the simplified (trimmed) fit curves. The area underneath the original load-displacement curves were calculated and plotted in blue color bars in Fig. 12.5b. The areas underneath the spline-fit curves are also plotted in red color and compared with the original curve areas. While the original curve areas of two specimens are approximately 10 % different because of the kink, the spline-fit curve areas are nearly identical.

By carefully inspecting the failed specimens, it was observed that the specimens failed in one of three failure modes depending on the displacement rates: cohesive failure mode (denoted as “C”) at relatively slow rates, interfacial failure mode (denoted as “I”) at relatively higher rates, and their mixed mode (denoted as “M”) at intermediate rates. Fig. 12.7a shows the microscopic image of the PCB surface of the cohesively (“C”) failed specimen, which was tested at a displacement rate of 0.0021 mm/s (0.005 in/min). The solder paste remains on the failed section uniformly. Fig. 12.7b shows the microscopic image

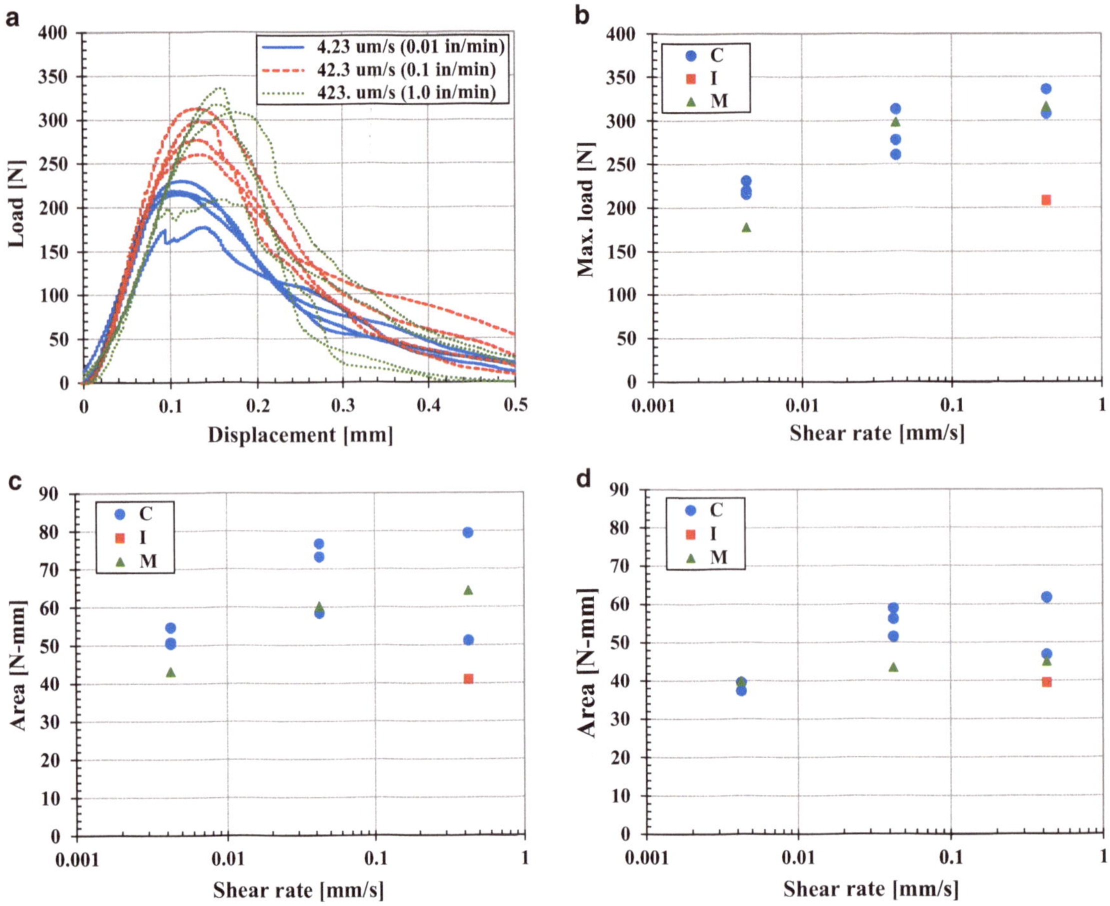


**Fig. 12.8** (**a**) Comparison of load-displacement curves under shear loading at three displacement rates for metallic Sn/Pb solder; (**b**) Maximum load of load-displacement curves against applied shear displacement rates; Area underneath (**c**) original and (**d**) trimmed load-displacement curves against applied shear displacement rates. "C", "I", and "M" represent cohesive solder failure, interfacial pad failure, and mixed cohesive and interfacial failure modes, respectively

of the PCB surface of the interfacially ("I") failed specimen, which was tested at a displacement rate of 2.12 mm/s (5.0 in/min). The failed section of tested specimen clearly indicates that failure occurred at the interface between the woven-fabric PCB and the Cu pad on the PCB surface. Fig. 12.7c shows the microscopic image of the PCB surface of the mixed ("M") failed specimen, which was tested at a displacement rate of 0.42 mm/s (1.0 in/min). The image shows a mixture of the cohesive and interfacial failure mode. Some portion of pad surface area is covered by the solder paste, while the rest of the area shows the interfacial peeling between the PCB and the Cu pad.

Twelve more specimens were fabricated with the Sn63/Pb37 solder. Among the 12 specimens, four specimens were tested at each of displacement rates: 4.23 μm/s (0.01 in/min), 42.3 μm/s (0.1 in/min), and 423 μm/s (1.0 in/min). The load-displacement curves obtained from the shear tests at the three displacement rates are plotted and compared in Fig. 12.8a. Blue, red and green curves are for the displacement rates of 4.23 μm/s (0.01 in/min), 42.3 μm/s (0.1 in/min), and 423 μm/s (1.0 in/min), respectively. The maximum loads are plotted against the three displacement rates in Fig. 12.8b. The calculated areas underneath the load-displacement curves, which are related to the interfacial shear-mode fracture toughness, are plotted against the applied shear displacement rates in Fig. 12.8c (area under original curve) and Fig. 12.8d (trimmed area by spline-fit curve). Marker labels, "C", "I" and "M", indicate the cohesive failure mode at the solder layer; the interfacial failure mode between PCB and Cu pad; and the mixed cohesive and interfacial failure mode, respectively. The general trend is that the solder failure is a dominant failure mode at low shear rates, while the interfacial and mixed failure modes become more frequent as the shear rate increases. Such strain-rate dependency damage responses coincide with the results from numerous previous studies that have shown that, with increased strain rate, the tendency for brittle (interfacial) fracture of solders increases [16–20]. Note that the "C" and "M" failure modes yield similar maximum load levels, and much higher than the "I" failure mode, as shown in Fig. 12.8b. However, "C" failure mode yields higher fracture toughness than the "I" and "M" failure modes, as shown in Fig. 12.8c, d. The "I" failure mode yields significantly smaller area than "C" and "M" failure modes. The areas calculated with

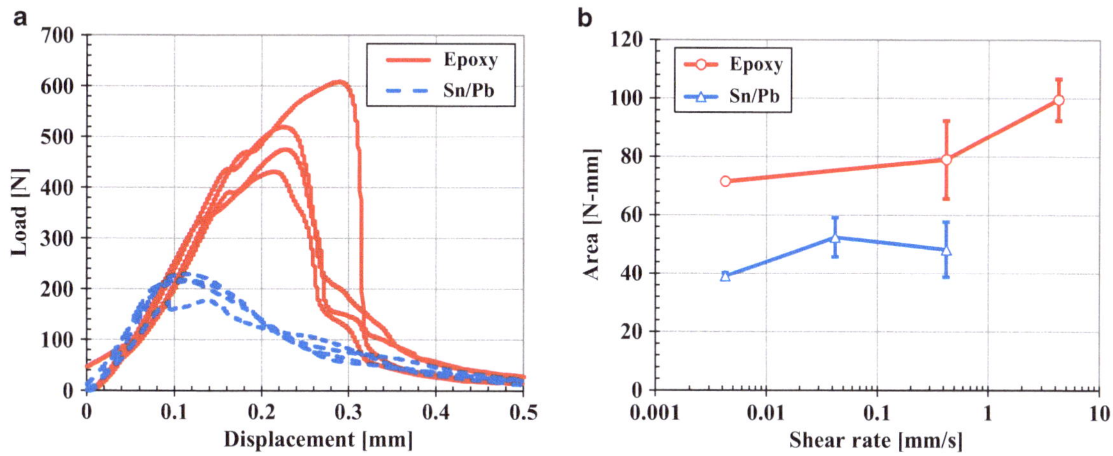


**Fig. 12.9** (**a**) Comparison of load-displacement curves of metallic (Sn/Pb) and polymeric (Epoxy) solders under shear loading at a displacement rate of 0.254 mm/min (0.01 in/min); (**b**) Area underneath load-displacement curves against various shear displacement rates

the spline-fit curves yields less scattered data than those from the original curves. The overall trend is that both maximum load and fracture toughness increase significantly with the increase of the displacement rates.

Polymeric solder materials have been developed to enhance the damage resistance under the impact loading. Because of their dielectric nature, polymeric materials need to be mixed with fillers having high electrical conductivity and percolation threshold. Possible filler materials with high conductivity are carbon nanotubes, carbon black, etc. Several shear test specimens were fabricated with epoxy material as solder. In this study, we used neat Epon 828 epoxy resin without any fillers. The epoxy was cured with the curing agent Jeffamine D400. Fig. 12.9a shows the load-displacement curves of the epoxy solder specimens, and compares them with those of the metallic Sn/Pb solder specimens. The displacement rate of these tests is 0.254 mm/min (0.01 in/min). Although initial slope of the curves are similar for both types of solder specimens, the maximum load and area underneath the curves of the epoxy solder specimens are significantly larger than the Sn/Pb solder specimens. Fig. 12.9b compares the areas underneath the curves of the epoxy and the Sn/Pb solder specimens obtained from various displacement rate tests, and clearly shows that the areas of epoxy solder specimens are significantly larger than those of the metallic solder for wide ranges of load rates. Therefore, the epoxy solder can be a stronger and tougher material, and thus enhance the survivability under the shear fracture dominant load. Note that this conclusion is only valid for the relatively low loading rate. Tests at higher rates will be conducted in the future study.

## 12.4 Summary and Conclusions

A new experimental method was developed to characterize critical interfacial damage parameters of solder interconnects subjected to various strain-rate mechanical loading, ultimately simulating shock or impact loading. The test apparatus and specimen were devised to experimentally characterize such critical damage parameters, particularly interfacial shear strength and fracture toughness at various strain rates. The test apparatus was designed to easily mount and unmount test specimens, and accommodate various sizes of components and solder layer thicknesses. Test specimens were fabricated with both metallic and polymeric solder materials, and tests were conducted under various shear displacement rates. Maximum load and area as the measure of fracture toughness were characterized from measured load-displacement curves. Three failure modes depending on the load rates were identified from optical microscopy. It was found that both strength and fracture toughness exhibit significant rate-dependency. Computational simulations were also conducted to correlate the measured area of the curves with the interfacial fracture energy, which will further be used for the damage prediction of the electronic components under the high-g impact loading.

**Acknowledgments** This work was financially supported by U.S. Air Force Research Laboratory with Contract No. FA8650-10-D-5011. The authors also greatly appreciate Arthur J. Safriet, Dayle Y. Pearson, Jacob Iarve, Chenggang Chen, Adam R. Waite, John D. Camping, Marlene D. Houtz, Albert M. Hilton, and Jeff L. Brown for providing valuable sample preparation, support and discussion.

## References

1. JEDEC Standard, JESD22-B111: Board Level Drop Test Method of Components for Handheld Electronic Products (2003)
2. JEDEC Standard, JESD22-B110A: Subassembly Mechanical Shock (2003)
3. JEDEC Standard, JESD22-B104C: Mechanical Shock (2004)
4. Lai, Y.-S., Yang, P.-F., Yeh, C.-L.: Experimental studies of board-level reliability of chip-scale packages subjected to JEDEC drop test condition. Microelectron. Reliab. **46**(2–4), 645–650 (2006)
5. Luan, J.-E., Tee, T.Y., Pek, E., Lim, C.-T., Zhong, Z., Zhou, J.: Advanced numerical and experimental techniques for analysis of dynamic responses and solder joint reliability during drop impact. IEEE Trans. Compon. Packag. Technol. **29**(3), 449–456 (2006)
6. Lall, P., Gupta, P., Panchagade, D., Angral, A.: Fault-isolation in portable electronics subjected to drop and shock. 12th IEEE Intersociety Conference on Thermal and Thermomechanical Phenomena in Electronic Systems (ITherm), 2–5 June 2010
7. Kaplan, S.M., Greendyke, R.B.: Finite element modeling of dynamic circuit board strain response under high-G shock impact. J. SMT **28** (4) (2015)
8. Lall, P., Gupte, S., Choudhary, P. Suhling, J.: Solder-joint reliability in electronics under shock and vibration using explicit finite-element sub-modeling. The 56th Electronic Components and Technology Conference, 2006
9. An, T., Qin, F.: Fracture simulation of solder joint interface by cohesive zone model. International Conference on Electronic Packaging Technology and High Density Packaging (ICEPT-HDP), 10–13 Aug 2009
10. An, T., Qin, F.: Cracking of the intermetallic compound layer in solder joints under drop impact loading. J. Electron. Packag. **133**(3), 031004–031004-7 (2011)
11. Bo, L.Z., Keat, L.W., Tamin, M.N.: Damage mechanics of solder/IMC interface fracture in Pb-free solder interconnects. The 11th Electronics Packaging Technology Conference (EPTC), 9–11 Dec 2009
12. Caers, J.F.J.M., Zhao, X.J., Wong, E.H., Seah, S.K.W., Selvanayagam, C.S., van Driel, W.D., Owens, N., Leoni, M., Tan, L.C., Eu, P.L., Yi-Shao, L., Chang-Lin, Y.: A study of crack propagation in Pb-free solder joints. IEEE Trans. Electron. Packag. Manuf. **33**(2), 84–90 (2010)
13. Gao, F., Jing, J., Liang, F.Z., Williams, R.L., Qu, J.: Cohesive zone simulation on dynamic fracture at the interfaces of a single solder joint. The 10th International Conference on Thermal, Mechanical and Multi-Physics simulation and Experiments in Microelectronics and Microsystems, 26–29 Apr 2009
14. Yamin, A.M., Shaffiar, N.M., Loh, W.K., Tamin, M.N.: Extended cohesive zone model for simulation of solder/IMC interface cyclic damage process in Pb-free solder interconnects. The 35th IEEE/CPMT International Electronic Manufacturing Technology Symposium (IEMT), 6–8 Nov 2012
15. Zhang, Y., Xu, Y., Liu, Y., Schoenberg, A.: The experimental and numerical investigation on shear behaviour of solder ball in a wafer level chip scale package. The 60th Electronic Components and Technology Conference (ECTC), 1–4 June 2010
16. Darveaux, R., Reichman, C.: Ductile-to-brittle transition strain rate. The 8th Electronics Packaging Technology Conference (EPTC), 6–8 Dec 2006
17. Qin, F., An, T., Chen, N.: Strain rate effects and rate-dependent constitutive models of lead-based and lead-free solders. J. Appl. Mech. **77**(1), 011008–011008-11 (2009)
18. Wang, X.J., Wang, Z.G., Shang, J.K.: Effect of displacement rate on lap shear test of SAC solder ball joints. International Conference on Electronic Packaging Technology and High Density Packaging (ICEPT-HDP), 28–31 July 2008
19. Yeh, C.-L., Lai, Y.-S.: Strain-rate and impact velocity effects on joint adhesion strength. International Conference on Electronic Packaging Technology and High Density Packaging (ICEPT-HDP), 28–31 July 2008
20. Frear, D.R., Vianco, P.T.: Intermetallic growth and mechanical behavior of low and high melting temperature solder alloys. Metall. Mater. Trans. A **25**(7), 1509–1523 (1994)

# Chapter 13
# Dynamic Mode II Delamination in Through Thickness Reinforced Composites

**Mehdi Yasaee, Galal Mohamed, Antonio Pellegrino, Nik Petrinic, and Stephen R. Hallett**

**Abstract** Through thickness reinforcement (TTR) technologies have been shown to provide effective delamination resistance for laminated composite materials. The addition of this reinforcement allows for the design of highly damage tolerant composite structures, specifically when subjected to impact events. The aim of this investigation was to understand the delamination resistance of Z-pinned composites when subjected to increasing strain rates.

Z-pinned laminated composites were manufactured and tested using three point end notched flexure (3ENF) specimens subjected to increasing loading rates from quasi-static (~0 m/s) to high velocity impact (5 m/s), using a range of test equipment including drop weight impact tower and a split Hopkinson bar (SHPB).

Using a high speed impact camera and frame by frame pixel tracking of the strain rates, delamination velocities as well as the apparent fracture toughness of the Z-pinned laminates were measured and analysed. Experimental results indicate that there is a transition in the failure morphology of the Z-pinned laminates from quasi-static to high strain rates. The fundamental physical mechanisms that generate this transition are discussed.

**Keywords** Through thickness reinforcement • Damage tolerance • Delamination testing • Z-pins • Strain rate • Impact

## 13.1 Introduction

Laminated composite materials have seen an increased usage across all transport sectors. The shift to use this lightweight material is necessary to help meet global environmental standards as well as reduce operational cost and improve performance. Composite materials provide exceptional specific stiffness and strength relative to their metal counterparts. Furthermore, they also possess excellent corrosive resistance and fatigue performance, yet a major drawback of this material is their lack of through thickness strength, which often results in disbond of the inter-laminar layers, also know as delamination. Composite materials are by design, capable of carrying in-service in-plane loads, however localized out of plane loading in forms of impact will result in delamination damage, which significantly reduces the performance of the component.

Many technologies have been developed over the years to address this problem, with through thickness reinforcement (TTR) being one of the highly successful methods [1]. A process known as Z-pinning is a popular method used to reinforce pre-preg composite laminates. Z-pining consists of inserting fibres or small rods into the composite material, reinforcing the thickness direction of the laminate. These small stiff, fibrous composite rods in the thickness direction bridge any newly formed delamination damage leading to excellent damage resistance capability [2].

Resistance of TTR composites to delamination has been subject to many studies, including quasi-static [2–4] and fatigue loading [5]. However, experimental investigations on the response of TTR composites when subjected to dynamic loading is limited and not well understood.

It has been shown for TTR composite laminates to have a direct dependence on strain rate when measuring the apparent mode II fracture toughness. Colin de Verdiere et al. [6] reported a modest increase of approximately 26 % in the initiation

M. Yasaee (✉)
School of Aerospace, Transport and Manufacturing, University of Cranfield, Cranfield MK43 0AL, UK
e-mail: m.yasaee@cranfield.ac.uk

G. Mohamed • S.R. Hallett
Advanced Composites Centre for Innovation and Science (ACCIS), University of Bristol, Aerospace Engineering, Queen's Building, University Walk, Bristol BS8 1TR, UK

A. Pellegrino • N. Petrinic
Department of Engineering Science, Engineering and Technology Building, Oxford University, Parks Road, Oxford, OX1 3PJ, UK



A.T. Zehnder et al. (eds.), *Fracture, Fatigue, Failure and Damage Evolution, Volume 8*,
Conference Proceedings of the Society for Experimental Mechanics Series, DOI 10.1007/978-3-319-42195-7_13

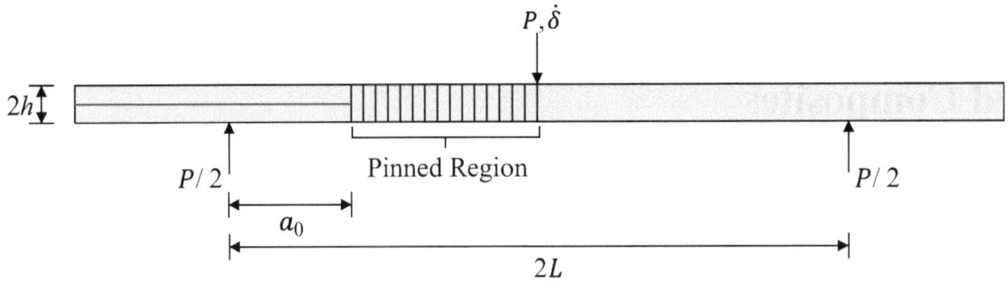


**Fig. 13.1** 3ENF test setup

apparent $G_{IIC}$ of tufted composite specimens. For Z-pinned composites the mode I apparent fracture toughness appears to reduce with an increase in loading rate as shown by Liu et al. [7].

Very little literature exists on the mechanical properties and the delamination resistance of Z-pinned composites' dependence on strain rate (e.g. [8]). The aim of this investigation was to carry out a set of controlled mode II fracture toughness tests of a laminated composite reinforced in the thickness direction using carbon fibre reinforced plastic (CFRP) rods or Z-pins. These tests were carried out at loading rates from quasi-static up to 5 m/s and full analysis of the composite response was made to conclusively show the effect of strain rate on the delamination resistance in un-reinforced and TTR epoxy based composites.

## 13.2 Materials and Specimen Preparation

Specimens were manufactured using IM7/8552 prepreg (Hexcel, UK) stacked with a sequence of $\left[(0, -45, 0, +45)_{3S}\right]_S$to achieve a nominal thickness of 6 mm. A 13 μm PTFE film was placed at the mid plane to act as an artificial starter crack. The test procedure followed the standard 3 point bend end notched flexure (3ENF) [9] shown in Fig. 13.1 with varying loading displacement rates ($\dot{\delta}$).

The Z-pinned specimens where pinned with T300 carbon/BMI pins arranged in a grid pattern with a spacing of 1.75 mm, generating a nominal 2 % areal density. To ensure consistency across specimens for both the control and the Z-pinned samples they were machined from a single plate.

Each specimen was machined to a nominal width of 20 mm. The un-cracked part of each individual specimen was tested in 3 point bending (3PB) following the ASTM-790 [10] test standard to measure the flexural modulus ($E_{1f}$) of the material. The width ($B$) and thickness ($2h$) of each specimen was measured at three different locations along its length to an accuracy of ±0.05 mm. For each specimen, a natural mode II pre-crack from the starter film was created using the procedure set out in ASTM-D7905 [9] to generate an initial crack length ($a_0$) of 20 mm when positioned in the final test configuration. This resulted in 30 mm of uncracked laminate and reinforced region ahead of the crack for the control and Z-pinned samples respectively.

## 13.3 Test Procedures

Increasing displacement loading rates from quasi-static ($8.3 \times 10^{-6}$ m/s), to intermediate (1–4 m/s) and high (5.5 m/s) where carried out on three different test apparatus. The support roller half span ($L$) was set at 50 mm with an initial crack length ($a_0$) of 20 mm and support roller and loading nose diameter of 10 mm were set for all the tests. Using high speed photography with a minimum of 100,000 fps for high loading rate tests and high definition imaging for quasi-static tests the displacement and the crack propagation for all tests were monitored.

Each camera was set up to ensure on average a 12 pixel per mm resolution, sufficient for full field strain measurements.

The quasi-static 3ENF tests were carried out according to the ASTM-D7905 [9] standard with a loading displacement rates of 0.5 mm/min ($8.3 \times 10^{-6}$ m/s). The load was measured using a calibrated 5 kN load cell on a hydraulic Instron test machine. For these tests, the instability in the initiation of delamination means the maximum load corresponded to the critical load.

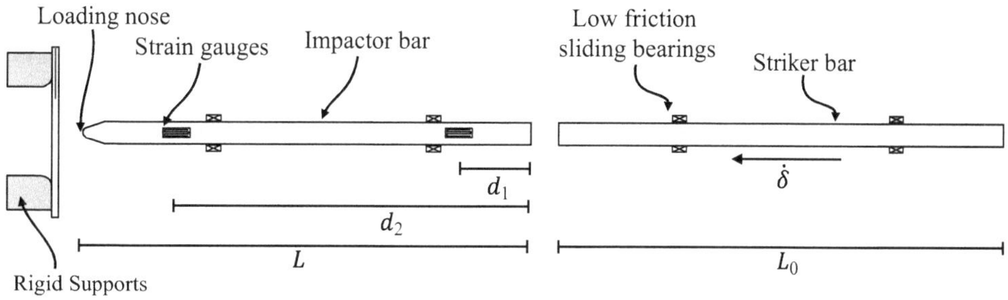


**Fig. 13.2** SHPB test setup

Intermediate loading displacement rate 3ENF tests were carried out on an instrumented drop weight impact tower. For these tests a cylindrical loading nose was attached to the end of a calibrated piezo-electric load-cell. The loading displacement rate was varied by raising the entire impactor unit, weighing 6.21 kg, to a specific height above the top surface of the laminate.

High loading displacement rate 3ENF tests were carried out using a modified Split Hopkinson Bar (SHPB) test procedure shown in Fig. 13.2. This experimental procedure is similar to the impact bending test procedure carried out by Hallett [11]. A striker bar of length $L_0$, is accelerated using compressed air to strike an instrumented impactor bar of length $L$ with the same mechanical impedance and diameter. This impact then generates an impact duration of $2L_0/c$, where $c = \sqrt{E/\rho}$ is the longitudinal wave speed in the bar. This transfer of kinetic energy then accelerates the impactor bar to a specific impact velocity generating the loading rate required to impact the specimen.

## 13.4 Data Reduction Technique

High frequency oscillations due to dynamic effects are inherent in all high rate tests. These dynamic effects are amplified as the loading rate increases therefore accurate critical load measurements are not possible [12] which will yield incorrect values of the materials fracture toughness. An alternative method proposed by Blackman et al. [12, 13] use a compliance based approach to calculating the $G_C$ of the material since the CFRP laminates exhibit no observable strain rate dependency in their axial modulus $E_{11}$.

For this investigation $G_{IIC}$ is calculated using the displacement at the moment of delamination initiation. This displacement can be reliably measured using the high speed photography images from all the different loading rate test procedures. The compliance of the 3ENF specimen [14] is given by:

$$C = \frac{2L^3 + 3a^3}{8E_{1f}Bh^3} + \frac{3L}{10G_{13}Bh} \tag{13.1}$$

The term on the right includes the influence of through thickness shear which is dependent on the $h/L$ of the test setup. The mode II inter-laminar fracture toughness is then calculated using [15]:

$$G_{IIC} = \frac{9\left(\frac{\delta}{C}\right)^2 (a + 0.42\chi h)^2}{16B^2 E_{1f} h^3} \tag{13.2}$$

$$\chi = \left[\frac{E_{11}}{11G_{13}}\left(3 - 2\left(\frac{\Gamma}{1+\Gamma}\right)^2\right]^{1/2} \tag{13.3}$$

$$\Gamma = \frac{1.18\sqrt{E_{11}E_{33}}}{G_{13}} \tag{13.4}$$

Where the term $0.42\chi h$ is the correction added to the length of the crack to account for the root rotation of the beam arms [15] and $E_{1f}$ is the flexural modulus of the material which was measured for each specimen independently in the current experiments.

## 13.5 Results

### 13.5.1 *Quasi-Static: Data Reduction Method Comparison*

The load-displacement plot of the control and pinned samples is shown in Fig. 13.3a. The quasi-static flexural tests of all the samples produced an average flexural modulus, $E_{1f}$ of 83.5 ± 1.1 GPa. In Fig. 13.3 the theoretical compliance using this flexural modulus average is plotted with $a = 20$ mm, $B = 20$ mm. Following the standard ASTM 3ENF test procedure the fracture toughness or $G_{IIC}$ of the IM7/8552 was calculated to be 663 ± 100 J/m$^2$. Using the compliance procedure described in Sect. 13.4 and Eq. (13.2) the fracture toughness was measured to be 673 ± 112 J/m$^2$. With only 1.5 % difference between the two procedures, the compliance procedure can be accepted to produce correct values of the fracture toughness of the material and gives confidence to use for the high rate procedure.

The average R curve for the control and pinned samples are shown in Fig. 13.3b. The 3ENF tests typically produce a single critical strain energy release rate value due to the unstable nature of the crack. However, with the addition of TTR the crack is arrested and thus propagates stably in the TTR region. This produces an increasing R curve with crack length due to the development of the extrinsic bridging zone behind the crack tip. There is a minor increase in the average critical strain energy release rate at the moment of initiation 922 ± 109 J/m$^2$ reaching a maximum strain energy release rate of 2613 ± 499 J/m$^2$ at a crack length of 50 mm. In this test configuration the maximum bridging zone length possible is 30 mm, however the fully developed Z-pin bridging zone length is expected to be much longer than the 30 mm length, approximately between 40 and 60 mm [16]. The apparent fracture toughness increase of these tests agrees well with that previously reported in literature [2, 16, 17].

### 13.5.2 *Load-Displacement Response*

The load-displacement plots for all the tests are given in Fig. 13.4. As expected, increase in displacement loading rate $\dot{\delta}$ increases the noise in the load output measured producing an unclear critical load prior to delamination. It can be seen that the critical load cannot be taken directly from the load displacement responses necessitating the use of the compliance procedure to calculate the $G_{IIC}$ of the specimens.

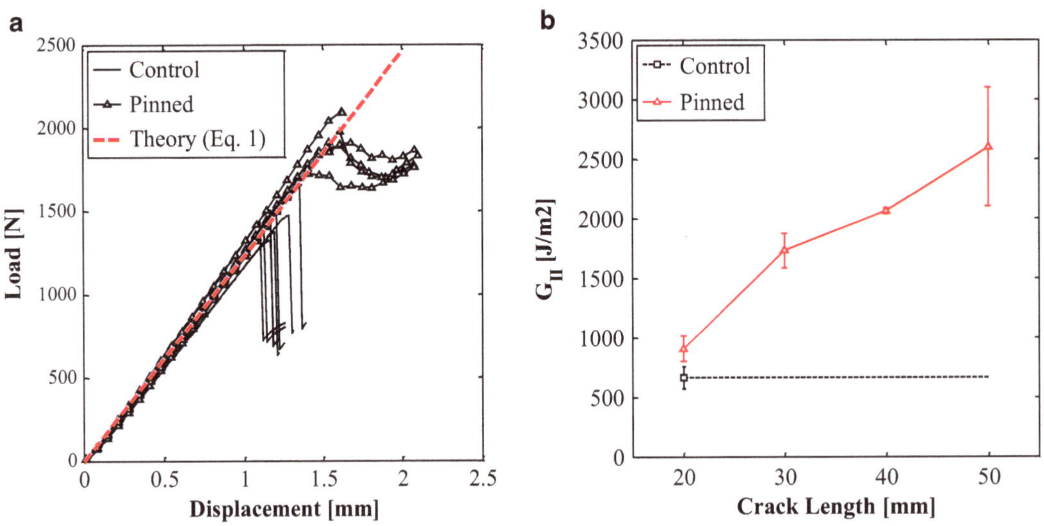


**Fig. 13.3** (**a**) Load-displacement for control specimens along with average compliance using Eq. (13.1). (**b**) Average R curve for control and pinned specimens

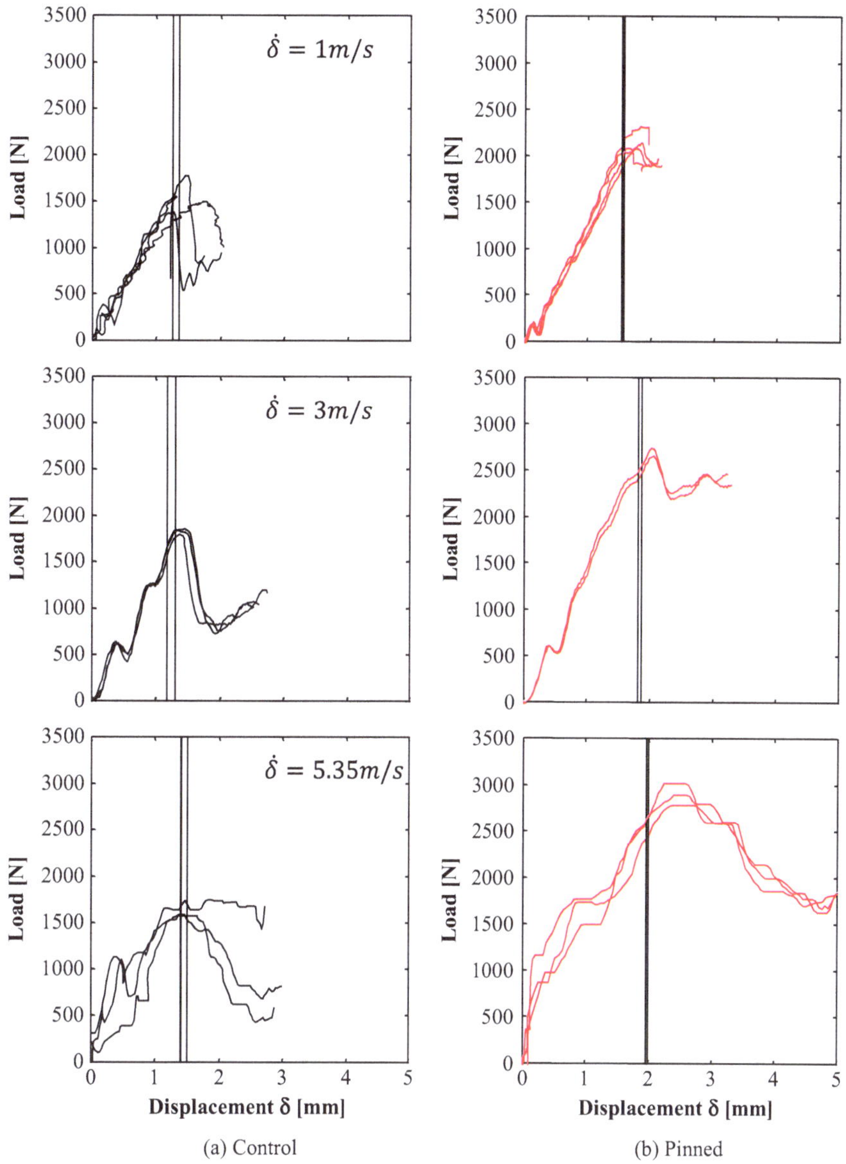


**Fig. 13.4** Load-displacement (δ) plots of for increasing loading displacement rate ($\dot{\delta}$), *dashed lines* indicate the displacement at which delamination initiated. (**a**) Control. (**b**) Pinned

The load response appears to be constant in the control samples with increasing $\dot{\delta}$. However for the pinned specimens the critical load appears to increase with $\dot{\delta}$. Furthermore, as expected, the load drop is distinctly sharper for the control samples, whereas the pinned specimens still maintain significant residual interlaminar strength after delamination initiation.

### *13.5.3 Rate Dependence of Interlaminar Fracture Toughness*

The calculated $G_{IIC}$ at the moment of delamination initiation against loading displacement rate ($\dot{\delta}$) is presented in Fig. 13.5. The control samples produce a minor increase in the $G_{IIC}$ with increase in loading rate, from 663 ± 100 J/m$^2$ for quasi-static tests to 970 ± 90 J/m$^2$ for $\dot{\delta}$ of 5.3 m/s. The pinned samples showed a very strong increase in $G_{IIC}$ with increase in loading rate. With initiation $G_{IIC}$ of 922 ± 109 J/m$^2$ for quasi-static tests to 2002 ± 64 J/m$^2$ for $\dot{\delta}$ of 5.3 m/s.

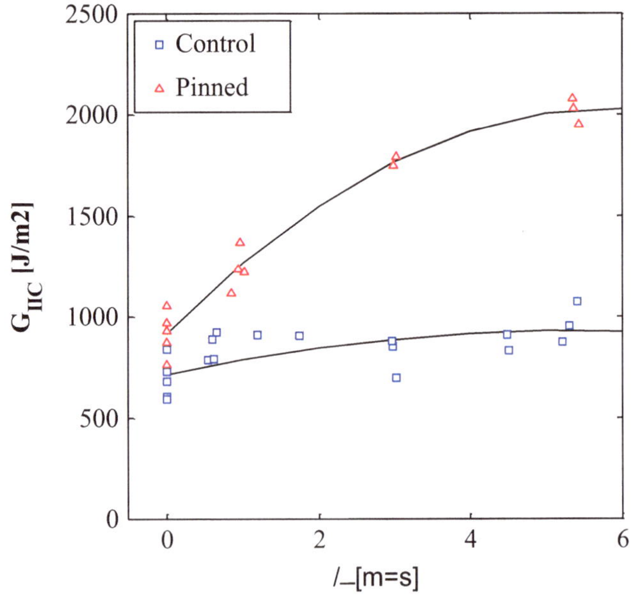


**Fig. 13.5** $G_{IIC}$ plots of for increasing loading displacement rate ($\dot{\delta}$)

## 13.6 Conclusions

An experimental characterisation of a mode II delamination in a Z-pin reinforced and unreinforced laminated composite has been carried out with increasing strain rates. Tests were performed on standard hydraulic test machines for quasi-static tests, instrumented drop-weight impact tower for intermediate loading rates and a SHPB for high loading rates. Using a compliance based approach rather that the standard load based data reduction techniques, the $G_{IIC}$ of the material was calculated at increasing strain rates.

The range of loading rates used in this investigation was from quasi-static to ~5.3 m/s. The mode II fracture toughness of the composite was seen to have a minor increase from 663 ± 100 to 970 ± 90 J/m$^2$ confirming behaviours observed in literature of tests on thermosetting brittle epoxy composites, where either minor or no significant increase in $G_{IIC}$ were reported [12, 18].

In contrast, with the presence of the Z-Pins, the mode II apparent fracture toughness showed a significant dependence on the displacement loading rate. It was shown that the initiation $G_{IIC}$ increases from 922 to 2002 J/m$^2$ over the velocity range tested here.

This behavior may be attributed to the increased load transfer from the bulk material to the individual pins. The increased loading rate may harden the matrix and the Z-pin resin due to viscoelastic effects and thus the compliance of the pins is decreased. What this means, is that the length over which the Z-pins provide traction to the delaminated surfaces is shortened, meaning more Z-pins contribute to the fractured energy calculated at the initial crack length. It is expected that this contribution of fracture energy would reach an upper plateau with increasing strain rates, hence it would be important to experimentally characterize even higher rates in the future. However, with increasing loading rates, the influence of the kinetic energy of the specimen on the apparent fracture toughness calculations will become more significant and will have to be fully considered.

**Acknowledgements** The authors would like to acknowledge Rolls-Royce plc for their support of this research through the Composites University Technology Centre (UTC) at the University of Bristol, UK.

## References

1. Yasaee, M., Mohamed, G., Allegri, G., Hallett, SR.: Delamination resistance of through thickness reinforced composites. In: Proceedings of the 16th European Conference on Composite Materials, Seville (2014)
2. Partridge, I.K., Yasaee, M., Allegri, G., Lander, J.K.: Damage-tolerant composite structures by Z-pinning. In: Toughening mechanisms in composite materials, pp. 161–189. Elsevier, Burlington (2015). doi:10.1016/B978-1-78242-279-2.00006-8

3. Lenzi, F., Riccio, A., Clarke, A., Creemers, R.: Coupon tests on z-pinned and unpinned composite samples for damage resistant applications. Macromol. Symp. **247**, 230–237 (2007)
4. Partridge, I.K., Cartie, D.D.R.: Delamination resistant laminates by Z-Fiber pinning: part I manufacture and fracture performance. Compos. A: Appl. Sci. Manuf. **36**(1), 55–64 (2005)
5. Pegorin, F., Pingkarawat, K., Mouritz, A.P.: Comparative study of the mode I and mode II delamination fatigue properties of z-pinned aircraft composites. Mater. Des. **65**, 139–146 (2014)
6. Colin de Verdiere, M., Skordos, A.A., Walton, A.C., May, M.: Influence of loading rate on the delamination response of untufted and tufted carbon epoxy non-crimp fabric composites/Mode II. Eng. Fract. Mech. **96**, 1–10 (2012)
7. Liu, H., Yan, W., Yu, X., Mai, Y.: Experimental study on effect of loading rate on mode I delamination of z-pin reinforced laminates. Compos. Sci. Technol. **67**(7–8), 1294–1301 (2007)
8. Schlueter, A., Parab, N.D., Chen, W.: Loading rate effects on mode I delamination of Z-pinned composite laminates. In: Song, B., Casem, D., Kimberley, J. (eds.) Dynamic Behavior of Materials, Volume 1. Springer, Cham (2014)
9. ASTM-D7905-14. Standard test method for determination of the mode II interlaminar fracture toughness of unidirectional fiber-reinforced polymer matrix composites. ASTM Int. (2014)
10. ASTM D790-07. Standard test methods for flexural properties of unreinforced and reinforced plastics and electrical insulating materials. ASTM Int. 2007e1 (2007)
11. Hallett, S.R.: Three-point beam impact tests on T300/914 carbon-fibre composites. Compos. Sci. Technol. **60**(1), 115–124 (2000)
12. Blackman, B.R.K., Dear, J.P., Kinloch, A.J., Macgillivray, H., Wang, Y., Williams, J.G., Yayla, P.: The failure of fibre composites and adhesively bonded fibre composites under high rates of test—part I mode I loading-experimental studies. J. Mater. Sci. **30**, 5885–5900 (1995)
13. Weeks, C.A., Sun, C.T.: Modeling non-linear rate-dependent behavior in fiber-reinforced composites. Compos. Sci. Technol. **58**(3–4), 603–611 (1998)
14. de Moura, M.F.S.F., de Morais, A.B.: Equivalent crack based analyses of ENF and ELS tests. Eng. Fract. Mech. **75**, 2584–2596 (2008)
15. Wang, Y., Williams, J.G.: Corrections for mode II fracture toughness specimens of composites materials. Compos. Sci. Technol. **43**(3), 251–256 (1992)
16. Pegorin, F., Pingkarawat, K., Daynes, S., Mouritz, A.P.: Mode II interlaminar fatigue properties of z-pinned carbon fibre reinforced epoxy composites. Compos. A: Appl. Sci. Manuf. **67**, 8–15 (2014). doi:10.1016/j.compositesa.2014.08.008
17. Cartie, D.D.R., Troulis, M., Partridge, I.K.: Delamination of Z-pinned carbon fibre reinforced laminates. Compos. Sci. Technol. **66**, 855–861 (2006)
18. Compston, P., Jar, P.-Y.B., Burchill, P.J., Takahashi, K.: The effect of matrix toughness and loading rate on the mode-II interlaminar fracture toughness of glass-fibre/vinyl-ester composites. Compos. Sci. Technol. **61**(2), 321–333 (2001)

# Chapter 14
# Measurement of Bond Line Fracture Toughness in Adhesively Bonded Composite Structures by Nanoindentation

**Denizhan Yavas and Ashraf F. Bastawros**

**Abstract** This study explores a probable correlation between the degradation of bond line toughness in adhesively bonded joints and the degradation/changes in measurable mechanical properties of the adhesive layer by indentation hardness due to contamination. The proposed framework utilizes the large scale bridging of interfacial fracture proposed by Tvergaard and Hutchinson (Philos Mag A 70:641–656, 1994). A typical adhesive/adherend material system (EA9394/Hexcel IM7-G/8552) exposed to different contaminants at the same concentration was examined. Nano indentation technique, considered as a non-destructive testing methodology compared to the bond line thickness, was utilized to measure the adhesive mechanical properties. In addition, macroscopic mode-I fracture toughness was independently measured by double cantilever beam test. Finite element method employing cohesive zone method was used to rationalize the experimental results and the prospective scaling-laws. The combined experimental results of macroscopic properties and the numerical results of the interfacial properties suggest a scaling between the interfacial cohesive fracture toughness and the measurable flow stress. While the proposed scaling is verified to a common adhesive-adherend system in aerospace industry, with additional examination of other systems, the proposed scaling law facilitates the utilization of the non-destructively evaluated indentation hardness to serve as an indicator for the bond line macroscopic fracture toughness.

**Keywords** Composite materials • Adhesive joints • Adhesion • Contamination • Nano indentation • Bond line integrity

## 14.1 Introduction

Carbon fiber reinforced polymer (CFRP) composites have been extensively utilized as a major construction material in modern civil aircraft industry, because of their high strength-to-weight ratio. The challenge remains in the early detection of the interfacial progressive damage of bonded structures and the assessments of the current level of the effective interfacial fracture toughness, before the onset of a catastrophic failure. This study explores the prospective of a probable correlation between the contamination-induced degradation of bond line toughness in adhesively bonded joints and the degradation/changes in measurable mechanical properties of the adhesive layer, measured by indentation testing. Indentation testing can be thought of as a non-destructive evaluation procedure, wherein the residual indentation impression is of the same order as the surface roughness, and is much smaller than the thickness of the adhesive bond line. The proposed framework employs the large scale bridging of interfacial fracture proposed by Tvergaard and Hutchinson [1], hereafter denoted TH, wherein plasticity has a major role in the toughness of the polymeric adhesively bonded joints (see for example [2, 3]). In TH-framework, the effective work of fracture per unit area on the interface, $\Gamma_{eff}$ is a function of the interfacial work of separation, $\Gamma_0$ and the properties of the adhesive layer, including thickness h, yield strength $\sigma_y$ and strain hardening exponent, n. The TH-framework suggests a dimensionless correlation between $\Gamma_{eff}$ and, $\sigma_y$ among other interfacial parameters.

$\Gamma_{eff}$ can be written as the confluence of the interfacial cohesive (surface or binding) energy $\Gamma_0$ and the energy dissipation through plastic (or inelastic) deformation, $W_p$, such that

$$\Gamma_{eff} = \Gamma_0 + W_p \tag{14.1}$$

For a bond line thickness $h$ greater than $R_0$, which is a material-based length scale, TH showed that $\Gamma_{eff}$ becomes independent of the bond line thickness. $R_0$ is approximately the plastic zone size as long as $R_0 < h$. When $\Gamma_{eff} > \Gamma_0$ for large scale

D. Yavas • A.F. Bastawros (✉)
Department of Aerospace Engineering, Iowa State University, Ames, IA 50011-2271, USA
e-mail: bastaw@iastate.edu

 
A.T. Zehnder et al. (eds.), *Fracture, Fatigue, Failure and Damage Evolution, Volume 8*,
Conference Proceedings of the Society for Experimental Mechanics Series, DOI 10.1007/978-3-319-42195-7_14

bridging, the plastic zone size scales with $R_0$. When plasticity has negligible contribution and the crack propagate at the interface at constant applied stress intensity factor, $K = K_0$; $R_0$ is defined as a reference length scale,

$$R_0 \equiv \frac{1}{3\pi}\left(\frac{K_0}{\sigma_y}\right)^2 = \frac{1}{3\pi(1-\nu^2)}\frac{E\ \Gamma_0}{\sigma_y^2}. \tag{14.2}$$

Here, the Young's modulus $E$ and Poisson's ratio $\nu$ are the properties of the bond line material. For a wide range of material properties TH showed that for adhesive layer thickness $h/R_0 > 4$, full plasticity enhancement of toughness is reached.

In the proposed experimental framework, we will be looking for a similar correlation, wherein $\Gamma_{eff} \sim 1/\sigma_y^m$ for the practical range of $0.5 \leq \hat{\sigma}/\sigma_y \leq 2$ observed in the current degradation study. To explore such relation, $\Gamma_{eff}$ and $\sigma_y$ should be independently measured. Here, the Tabor approximation [4, 5] is being utilized for the indentation hardness to be considered as an average measure of the material yield strength in compression ($H \sim 3\sigma_y$). $\Gamma_{eff}$ will be measured experimentally from double cantilever beam (DCB) tests (ASTM D5528-13). The intrinsic fracture toughness properties will be examined numerically to match the experimental measurements $\Gamma_{eff}$ and the peak load of the force-displacement curve of the DCB sample. Once these correlations are established, we anticipate that a contamination induced softening within the adhesive layer will result in the increase of $\Gamma_{eff}$.

## 14.2 Material and Sample Preparation

Nanoindentation samples consist of a layer of Hysol EA9394 adhesive on top of a glass slide. These samples were exposed to different types of contaminants (Table 14.1), with controlled surface concentration of 55 μg/cm$^2$. The selected contaminants were first diluted by a volatile agent such as hexane in a proper concentration, then the diluted solution was applied uniformly on top of the uncured adhesive layer. The desired surface concentration was controlled by the height of the solution, and verified by performing infra-red spectroscopy to be within few percent. Finally the samples were cured in a furnace at 190 °F (88 °C) for 2 h according to Hysol EA9394 material data sheet.

The fracture DCB samples consist of two adherend composite panels bonded by a layer of adhesive. Adherend panels with a layup of $[\pm 45^o, 0_5^o]_S$, were constructed by stacking 10 plies unidirectional prepregs (Hexcel IM7-G/8552) and two additional surface plies of bi-directional weaved fabric (Hexcel W3B-282). Standard fabrication of vacuum bagging and hot-press curing methods were implemented. In order to obtain enhanced bonding, the adherend panel surfaces were treated by 220 alumina grit blasting under 60 psi (413.7 kPa) pressure and expected nozzle speed of 136 kg/h prior to the bonding. The resulting surface roughness of 4–5 μm was obtained, which is generally considered as sufficient surface roughness for a proper adhesion process. After the sand-blasting process, the panel surfaces were cleaned by water and hexane, then the panels were left in a drying furnace at 160 °F for 2 h with a heating and cooling rates of 5 °F/min, following NAVAIR 01-1A-21 (see Sec. 6-7a (2)) technical manual. The diluted contaminants were uniformly applied on the roughened and cleaned surface of the composite panels. The desired surface concentration was again verified by infrared spectroscopy measurements. Finally A pair of exposed/unexposed adherend panels were bonded to each other by an adhesive layer. A 125 μm thick Teflon film was used to create an initial crack at the edge of the samples. In addition, a 250 μm shim was inserted at the edges to control the adhesive layer thickness. The resulting bond line thickness was measured around 400 μm. Finally, bonded composite panels were cured under 5 psi (34.5 kPa) pressure at room temperature (22 °C) for 16 h. Subsequently, another final rapid curing process was performed in the furnace at 190 °F (88 °C) for 2 h. These assembled composite panels were cut into sample dimensions (25.4 × 180 mm$^2$) and aluminum loading blocks were inserted to the ends of the samples. Figure 14.1 shows the sample geometry and dimensions.

**Table 14.1** Different classes of contaminants used in this study

| | Notation | Class | Explanation |
|---|---|---|---|
| 1 | HF-1 | Aviation Hydraulic fluid | MIL-PRF-87257 |
| 2 | HF-2 | Aviation Hydraulic fluid | SAE AS1241 |
| 3 | HF-3 | Aviation Hydraulic fluid | MIL-PRF-85570 |
| 4 | F-44 | Mold release agent | LOCTITE FREKOTE 700-NC |
| 5 | F-700 | Mold release agent | LOCTITE FREKOTE 44-NC |

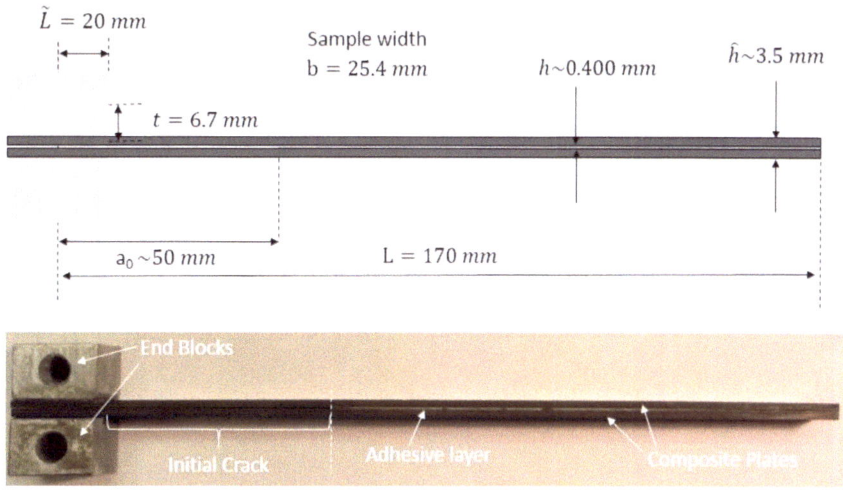


**Fig. 14.1** Double cantilever beam (DCB) test sample geometry and dimensions

## 14.3 Measurement of Adhesive Hardness by Nanoindentation

Nanoindentation technique was utilized to probe the changes in mechanical properties of the adhesive material due to contamination. All indentations were performed in force control using the Hysitron TI 950 TriboIndenter (TriboIndentor$^{TM}$ by Hysitron Inc.). A trapezoidal loading profile, which consists of a 5 s linear loading, 2 s hold at the peak load, and 5 s linear unload, was utilized to impose the regular patterns of nano-indentation using a Berkovich tip with a tip radius of about 100–150 nm. Several indentations were performed on each sample with a peak load ranging from 1000 to 6000 μN with 1000 μN intervals, which resulted in contact depths up to 5 μm. Figure 14.2a shows a representative set of force and indentation depth curves obtained from indentations preformed on the reference (unexposed), HF-1, and F-44 contaminated samples.

Figure 14.2b shows the variation in indentation-measured $H$ (derived from the contact analysis by Oliver and Pharr [6]) of the adhesive layer exposed to different types of contamination.. It should be noted that these measured values should not be mistaken with the bulk macroscopic values, especially the hardness as it might display a size dependent response at the nano-scale according to Hall-Petch strength-grain size relationship [7, 8]. Therefore, the percentage change with respect to the reference measurement are utilized in the comparison and also presented on Fig. 14.2b. Two distinct effects of contamination on $H$; namely softening and hardening were observed depending on the contaminant group. For the examined adhesive, $H$ is decreased by the hydraulic fluid contaminants (softening of the adhesive), whereas it is increased by the mold release agent contaminants (hardening of the adhesive). According to our hypothesis effective bond line toughness should scale with $H$ raised to a negative exponent, ($\Gamma_{eff} \sim 1/H^m$), when the interfacial cohesive strength remains higher than the adhesive yield strength ($\hat{\sigma} > \sigma_y$). Thus, nanoindentation results predict higher degradation in bond line toughness in case of contamination by mold release agents (accompanied with increase in hardness) compared to hydraulic fluids (accompanied with reduction in hardness). To verify such trend, DCB tests were performed to measure the corresponding bond line fracture toughness for each contaminant case, discussed in the following sections.

## 14.4 Measurement of Adhesive Fracture Energy by Double Cantilever Beam Test

Bond line fracture toughness was measured by DCB test, according to ASTM-D5528-13, which is a well-established testing method for mode-I fracture characterization of adhesively bonded structures. All DCB tests were performed in displacement control at a loading rate of 0.02 mm/s using an INSTRON 8862 servo-electric computer controlled testing frame. Force, cross-head displacement and in-situ digital images of the crack growth were recorded simultaneously. Three DCB tests were performed for each class of contaminants. A representative set of force-displacement curves obtained from testing of reference (unexposed), HF-1, HF-3, and F-44 contaminated samples is presented in Fig. 14.3a. The slight change in the initial loading stiffness arises from the differences in the initial crack length, $a_0$, being determined after an initial loading/unloading cycle to sharpen the manufactured interfacial crack-tip. The uncontaminated reference case has shown extended

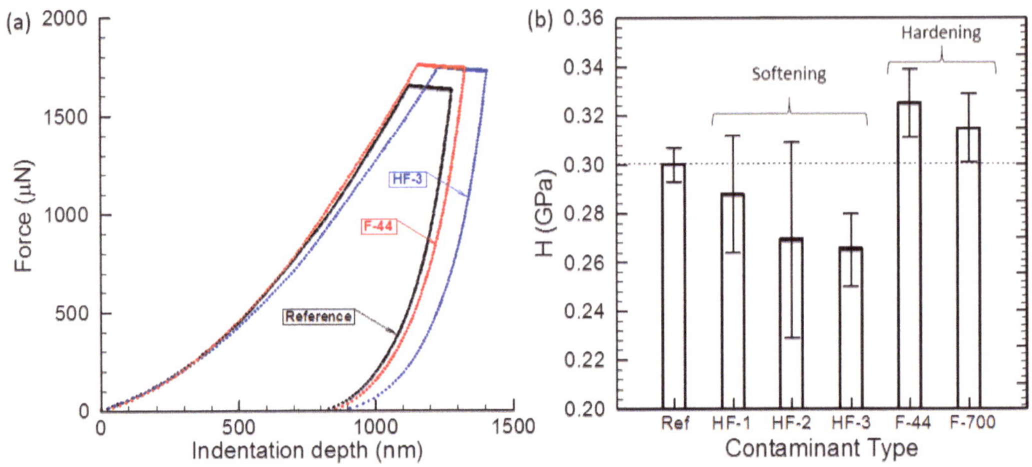


**Fig. 14.2** (**a**) Indentation force-depth curves for the samples exposed to different contaminants, (**b**) variation in Hardness (H) of the adhesive layer exposed to different contaminants

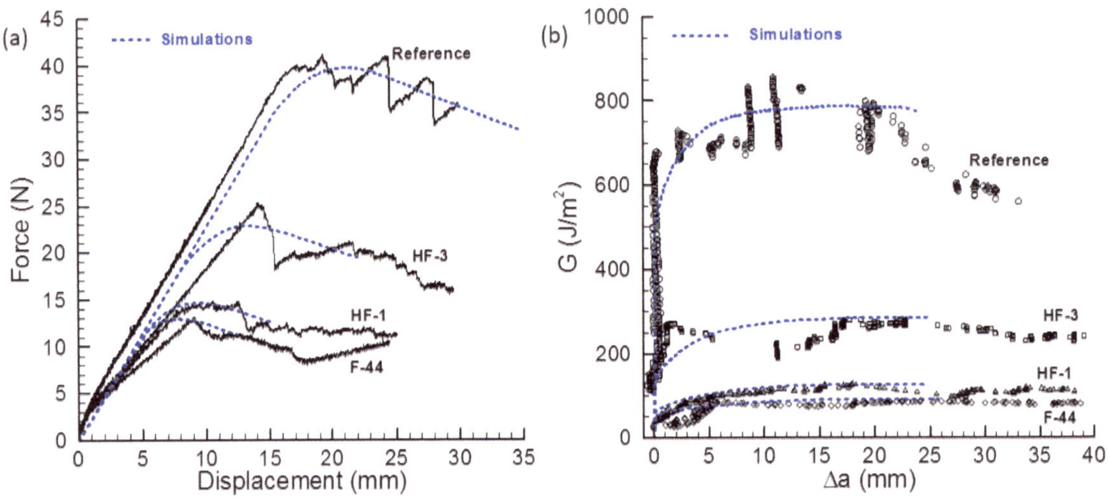


**Fig. 14.3** (**a**) Force-displacement curves and (**b**) corresponding fracture energy release rate curves, obtained from DCB test of reference (uncontaminated), HF-1, HF-3, and F-44 contaminants. The *dashed blue lines* are the results of the FEM cohesive zone simulations

nonlinearity on the force-displacement curve (about one third of the loading level) before the start of the crack propagation. To the contrary, all the contaminated cases have shown a linear rise up to the start of the crack propagation. This is the first evidence of the shielding of plasticity within the adhesive layer exposed to contamination. Upon reaching the maximum load carrying capacity of the joint, a progressive drop in the load is observed as a function of successive crack propagation. The associated zig-zag pattern is attributed to the intermittent propagation and arrest of the interfacial crack or relaxation of the fracture energy, followed by building up of the fracture driving forces. The corresponding derived mode-I energy release rate curves are presented in Fig. 14.3b, clearly showing the arrest and propagation of the crack for the reference case. However, the crack propagation is quite steady for the contaminated cases. This could be attributed to the absence of the plastic process zone within the bond line. In addition, it could be argued that propagation and arrest of the crack in the reference case is due to the reestablishment and collapse of the plastic process zone, within the bond line.

Figure 14.4a summarizes the average steady state levels of energy resale rate curves, which were considered as mode-I bond line fracture toughness. All contaminant cases show significant reduction of the adhesive fracture toughness. The different hydraulic fluid contaminants show a drop of about 4–6 times the reference value. The Frekote contaminants show more than an order of magnitude drop in the fracture toughness. For the reference case, the optical image insert shows a mixed-type failure between the adhesive and adherend, exposing the fibers within the adhesive layer. All other contaminant cases at the reported level of contamination show adhesive-type failure.

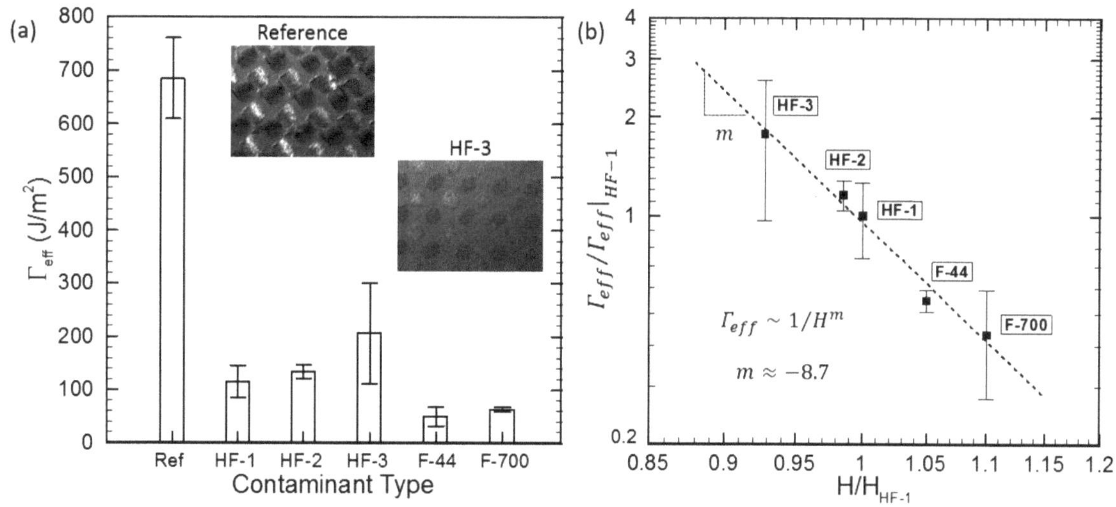


**Fig. 14.4** (**a**) Summary of Mode-I fracture toughness measurement for different contaminants (applied at the surface concentration of 55 μg/cm$^2$), highlighting the severe chemical degradation of the adhesive strength, inserts show the fractal surfaces, (**b**) correlation between DCB-test measured bond line fracture toughness and indentation hardness of bond line, where all the values are normalized by those of HF-1 contaminant and presented on logarithmic scale

The DCB fracture toughness results show some consistency with the hardness measurements as an overall trend of Fig. 14.2b. For example, the Frekote contaminants show increased hardness compared to the hydraulic fluid contaminants. Thus they should have lower level of fracture toughness compared to the reference and the hydraulic fluid contaminants. The hydraulic fluid contaminants show a drop in their hardness on Fig. 14.2b, which should have been manifested as an increase in the fracture toughness. To the contrary, the group of the hydraulic fluid contaminants shows drop in their fracture toughness compared to the reference case. Though, they are 2–5 times higher than the Frekote contaminants fracture toughness level. Additionally, the fracture toughness changes relative to each other in the same trend noted in Fig. 14.4a, though not with the same magnitude. To show this effect, Fig. 14.4b summarizes the normalized fracture toughness for all contaminant cases as a function of the normalized hardness on a logarithmic scale, with the values of HF-1 utilized for normalization. As can clearly be seen, hardness (or yield strength) has a strong influence on the measured bond line toughness. The correlation exponent is around $m \approx -8.7$. A similar correlation for the intrinsic fracture toughness properties will be examined numerically in the following sections.

## 14.5 Numerical Modeling and Scaling Law

Numerical simulations were performed using the commercial FEM package SIMULIA Abaqus 6.12-2. DCB sample geometry was modeled using 2-D 4-node quadrilateral linear plane strain elements (CPE4). A bilinear traction-separation law [9] was employed to model the bond line failure. Figure 14.5 shows the utilized bilinear traction-separation curve, which is composed of linear elastic deformation, followed by a linear softening domain representing the progression of crack surface displacement till full separation. In the linear elastic part, the cohesive surface tractions ($t_n$, $t_s$) and displacements ($\delta_n$, $\delta_s$) are related to each other by initial (penalty) stiffness matrix, **K**, where the cohesive surface will undergo linear elastic deformation until it reaches the critical traction values ($\hat{\sigma}$, $\hat{\tau}$). Thereafter, a linear softening will commence until reaching the critical separation values ($\hat{\delta}_n$, $\hat{\delta}_s$). Since DCB test is a pure mode-I fracture testing, a maximum nominal traction criterion ($t_n \geq \hat{\sigma}$ ; $\hat{\sigma}$ critical normal traction) is employed to initiate the damage. The cohesive fracture energy, $\Gamma_0$, which is the total area under the traction-separation curve, controls the damage evolution. In the current framework, complete surface separation commences when mode-I driving fracture forces reaches critical value, ($\Gamma \rightarrow \Gamma_0$) wherein $\Gamma_0$ is the mode-I cohesive fracture toughness.

Adherend composite panels were effectively modeled as an isotropic elastic material, whereas the adhesive layer was modeled as an isotropic elastic-plastic material. The effective flexural stiffness of the adherend material was derived from four point bend test according to ASTM-D7264/D7264M-07 and was estimated to be $E = 64$ GPa, $\nu = 0.3$. The mechanical properties of the adhesive were $E = 4.3$ GPa, $\sigma_y = 69$ MPa, $\nu = 0.43$, with modest hardening exponent of $n = 1/N \approx 6.2$. The two parameters $\Gamma_0$ and $\hat{\sigma}$ of the bilinear traction-separation law were calibrated by matching the peak force of the

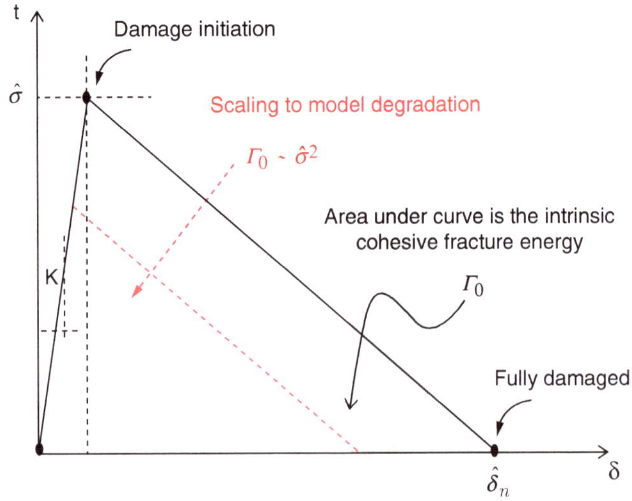


**Fig. 14.5** The utilized bilinear traction-separation law, showing self-similar scaling approach to model degradation

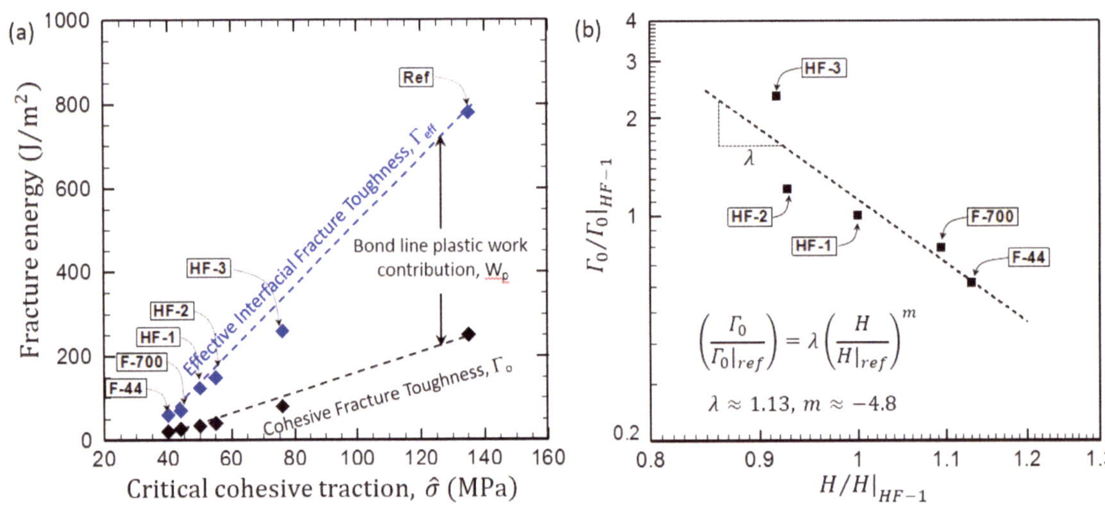


**Fig. 14.6** (**a**) Summary of the numerical results for the effective fracture toughness with sets of cohesive fracture toughness and critical cohesive traction, mapping the experimental cases. (**b**) Experimentally derived correlation between calibrated interfacial cohesive fracture toughness and measured hardness, where all the values are normalized by those of HF-1 contaminant and presented on logarithmic scale

experimentally measured force-displacement curve as proposed by Li et al. [10]. Then, $\Gamma_0$ was varied by scaling the linear softening domain while keeping its slope constant so that the traction-separation law remains self-similar as shown in Fig. 14.5. Such variability can impose the scalability between interfacial fracture toughness and interfacial cohesive strength, such that $\Gamma_0 \sim \hat{\sigma}^2$.

In the numerical simulations, the adhesive yield strength, $\sigma_y$ and modulus, $E$ were utilized from the nanoindentation measurements. The corresponding $\Gamma_0$ and $\hat{\sigma}$ were varied to match the experimentally measured force-displacement, and energy release rate-crack extension, of Fig. 14.3. The FEM results are the dashed blue lines on the figures. Qualitatively, the numerical simulations reproduced the experimentally measured trends in terms of matching maximum failure load and steady-state energy release rate values. Though the numerical prediction did not exhibit the intermittent propagation and arrest of the interfacial crack. To capture such effect, the analysis must account for the inertial effects as well as the stochastic distribution of defects, which are beyond the scope of the current study.

Figure 14.6a provides a numerically derived maps for the role of different parameters on the interfacial fracture toughness of the bond line. One can easily note the significant variation of the ratio $\Gamma_{eff}/\Gamma_0$ for each set of conditions (mapped to the mechanical properties of one of the contaminants). As a result, the amount of plastic work that could be dissipated within the adhesive layer is progressively deteriorating as well. It should be noted that $\Gamma_{eff}$ reaches $\Gamma_0$ for some of the worst contaminant cases, which is an indication of existence of very limited plasticity at the interface. Additionally, $\hat{\sigma}$ started to reach $\sigma_y$ and for the range of comparison it droped below $\sigma_y$. Accordingly the interface would decoher before the bond line can reach the state of plastic work dissipation. In other words, degradation of the critical cohesive traction due to contamination shields the plasticity within the adhesive material.

Furthermore, it could be inferred that $\Gamma_0$ scales with $W_p$ for the examined adherened/adhesive system and for the explored domain of experiment, from Eq. (14.1) and supported by the numerical results of Fig. 14.6a. Physically, $\Gamma_0$ should be controlled by the chemical interaction of the adhesive layer and the contamination. Though this correlation implies the possibility of $\Gamma_0$ scaling with $H$. Figure 14.6b explores such prospective correlation between the experimentally measured hardness and the numerically calculated interfacial work of adhesion $\Gamma_0$, normalized with the properties of contaminant HF-1. A strong correlation could be identified, clearly showing the effect of contamination on deteriorating the interface properties. Such correlation could be rationalized in a power law relationship such that,

$$\frac{\Gamma_0}{\Gamma_0|_{\text{Ref}}} = \lambda\left(\frac{H}{H|_{\text{Ref}}}\right)^m. \tag{14.3}$$

For the examined adhesive-adherend material system and for the diverse range of contaminants, $\lambda \approx 1.13$ and $m \approx -4.8$. In principle, correlation (Eq. 14.3) suggest that the measurement of $H$ for a contaminated case, and establishment of the macroscopic $\Gamma_{eff}$, and the intrinsic $\Gamma_0$ fracture properties of a reference case, one could project the effective fracture toughness of the contaminated case, for the range of bond line thickness $h/R_0 > 4$. Accordingly, the correlation given in Eq. (14.3) has the potential to be used for comparative study to map the role of contamination on the cohesive (or effective) fracture toughness on bonded composite structures.

## 14.6 Conclusions

The presented study showed the potential of nanoindentation technique to predict the bond line fracture toughness by monitoring changes in the measurable mechanical properties of the adhesive material due to contamination. A common aerospace adhesive-adherend (Hexcel IM7-G/8552-Hysol EA9394) material system was examined with exposure to different classes of environmental contaminants. The adhesive mechanical properties such as Young's modulus and hardness were measured by nanoindentation. Effective macroscopic mode-I fracture toughness was independently measured by double cantilever beam test. In addition, finite element method employing cohesive zone method was used to rationalize the experimental results and the prospective scaling-laws. A scaling between the interfacial cohesive fracture toughness and the measurable yield strength of the adhesive layer was suggested based on the combined experimental results of macroscopic properties and the numerical results. The proposed scaling law utilizes mechanical properties of the adhesive layer to elucidate the trends of the macroscopic fracture toughness for a common adhesive-adherend system in aerospace industry under different environmental exposures. With additional examination of other systems, the proposed scaling law facilitates the utilization of the non-destructively evaluated indentation hardness to serve as an indicator for the bond line macroscopic fracture toughness.

**Acknowledgements** This work was supported by NAVAIR through Redwood Scientific Inc. under contract No. 104728.

## References

1. Tvergaard, V., Hutchinson, J.W.: Toughness of an interface along a thin ductile layer joining elastic solids. Philos. Mag. A **70**(4), 641–656 (1994)
2. Chai, H.: On the correlation between the mode-I failure of adhesive joints and laminated composites. Eng. Fract. Mech. **24**(3), 413–431 (1986)
3. Chai, H.: Observation of deformation and damage at the tip of cracks in adhesive bonds loaded in shear and assessment of criterion for fracture. Int. J. Fract. **60**, 311–326 (1993)
4. Tabor, D.: Hardness of Metals. Oxford University Press, Oxford (1951)
5. Tabor, D.: Indentation hardness and its measurement: some cautionary comments. In: Blau, P.J., Lawn, B.R. (eds.) Microindentation Techniques in Materials Science and Engineering ASTM STP 889, pp. 129–159. American Society for Testing and Materials, Philadelphia (1986)
6. Oliver, W.C., Pharr, G.M.: An improved technique for determining hardness and elastic modulus using load and displacement sensing indentation experiments. J. Mater. Res. **7**(6), 1564–1583 (1992).

7. Hall, E.O.: The deformation and ageing of mild steel III. Proc. Phys. Soc. **64**, 747–753 (1951)
8. Cracknell, A., Petch, N.J.: Frictional forces on dislocation arrays at the lower yield point in iron. Acta Metall. **3**, 186–189 (1955)
9. Geubelle, P.H., Baylor, J.S.: Impact-induced delamination of composites: a 2D simulation. Compos. Part B **29B**, 589–602 (1998)
10. Li, S., Thouless, M.D., Waas, A.M., Schroeder, J.A., Zavattieri, P.D.: Use of mode-I cohesive-zone models to describe the fracture of an adhesively-bonded polymer-matrix composite. Compos. Sci. Technol. **65**, 281–293 (2005)

# Chapter 15
# Experimental and Numerical Investigation of Novel Crack Stopper Concepts for Lightweight Foam Cored Sandwich Structures

**W. Wang, G. Martakos, J.M. Dulieu-Barton, and O.T. Thomsen**

**Abstract** Three novel crack stopper designs for foam cored composite sandwich structures have been investigated with respect to their ability to deflect and arrest propagating face debond cracks. One of the new crack stoppers was similar to a previously developed design, whereas the two others were modified with layers of glass fibre fabric extending from the peel stopper tip into the face sheet, or into the face sheet/core interface. The novel designs were investigated under mode I dominated crack propagation conditions. Both quasi-static and fatigue loading scenarios were investigated. The mechanisms controlling crack propagation were studied using Thermoelastic Stress analysis (TSA) and Finite Element (FE) analysis. The TSA revealed significant new information about the local stress fields in the vicinity of the crack stopper tip as well as the fracture process zone. The first configuration in most cases was able to deflect debond cracks, albeit not in all cases, whereas it was incapable of achieving crack arrest. The two other designs performed better in that they consistently demonstrated the ability to deflect propagating cracks. Only the second design could arrest the cracks consistently as well. Detailed numerical fracture mechanics analyses confirmed and explained the experimental observations.

**Keywords** Foam cored composite sandwich structures • Peel stoppers • Thermoelastic stress analysis • Fracture modelling • Damage tolerance

## 15.1 Introduction

A well known weakness of sandwich structures is the bonding between the face sheet and core. To prevent the propagation of cracks from debonds at the interface of the face sheet and the core, 'crack stoppers' as shown in Fig. 15.1a may be introduced into the structure. The principle of the crack stopper is to deflect cracks away from the interface into the neighbourhood of the crack stopper wedge, and hence arrest its propagation. The present work studies the behaviour at the crack stopper wedge under mode I dominated loading. Thermoelastic stress analysis (TSA) is used to assess the stresses around the crack stopper during crack growth. The goal is to investigate local effects introduced by the crack stopper and to understand the fracture mechanism at the tri-material junction of the core, face sheet and crack stopper.

## 15.2 Experimental Investigation

As shown in Fig. 15.1a, the sandwich specimen consists of a PVC H100 foam core and glass/epoxy composite face sheets. The crack stopper was made from Permalock 9004 polyurethane (PU) adhesive. Three configurations of the crack stopper are studied. In configuration 1 (C1) the crack stopper is directly bonded to the foam; in configuration 2 (see C2 in Fig. 15.1b) a layer of fibre is moulded in the crack stopper and the fibre protruding from the crack stopper tip is attached to the face sheet.

W. Wang
Faculty of Engineering and the Environment, University of Southampton, Highfield SO17 1BJ, UK

G. Martakos
Department of Mechanical and Manufacturing Engineering, Aalborg University, 9220 Aalborg East, Denmark

Siemens Wind Power A/S, 9220 Aalborg East, Denmark

J.M. Dulieu-Barton • O.T. Thomsen (✉)
Faculty of Engineering and the Environment, University of Southampton, Highfield SO17 1BJ, UK

Department of Mechanical and Manufacturing Engineering, Aalborg University, 9220 Aalborg East, Denmark
e-mail: o.thomsen@soton.ac.uk



A.T. Zehnder et al. (eds.), *Fracture, Fatigue, Failure and Damage Evolution, Volume 8*,
Conference Proceedings of the Society for Experimental Mechanics Series, DOI 10.1007/978-3-319-42195-7_15

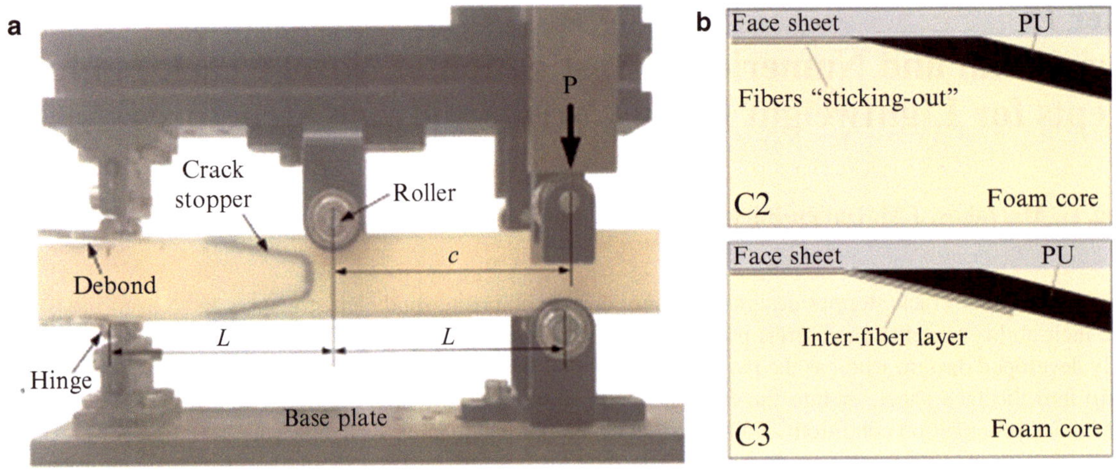


**Fig. 15.1** (**a**) The sandwich beam specimen loaded in MMB rig, (**b**) the configurations of the crack stopper

**Table 15.1** Crack growth paths for different specimens

| Configuration | Specimen | Crack path before the tri-material junction | Crack path at the tri-material junction |
|---|---|---|---|
| C1 | C1_f1 | Foam | Deflected |
| | C1_f2 | Foam | Not deflected |
| | C1_f3 | Foam | Not deflected |
| C2 | C2_f1 | Foam | Deflected |
| | C2_f2 | Foam | Deflected |
| | C2_f3 | Foam | Deflected |
| C3 | C3_f1 | Interface | Deflected |
| | C3_f2 | Interface | Deflected |
| | C3_f3 | Foam | Deflected |

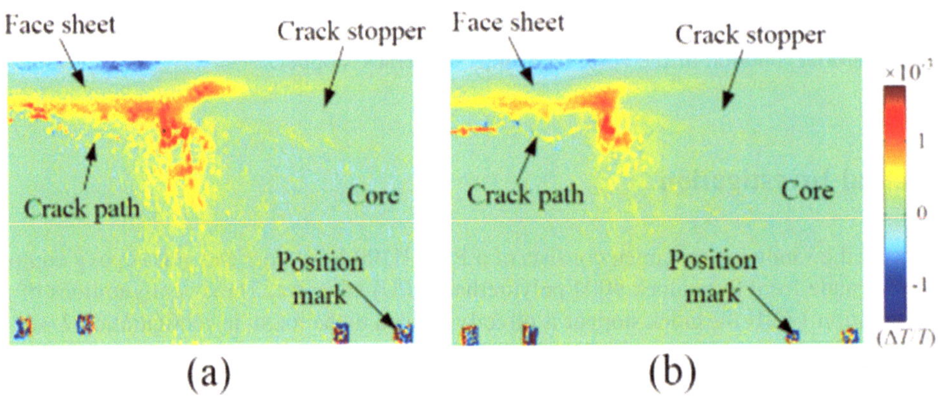


**Fig. 15.2** TSA image obtained at the tri-material junction obtained from (**a**) C1 (specimen C1_f1) and (**b**) C3 (specimen C3_f3)

Configuration 3 (C3) places a layer of fibre at the foam/PU interface and the fibre away from the crack stopper is attached to the face sheet. The specimen is loaded in a mixed mode bending rig (see Fig. 15.1a). A long lever arm distance, *c*, is used to provide the mode I dominated loading.

The sandwich specimens were tested under fatigue loading; the crack growth paths are shown in Table 15.1. For the C1 specimens, in two cases the crack was not deflected. A common observation from the C1 specimens is that the crack at the tri-material junction shows a clear trend to kink back to the face sheet. Figure 15.2 shows the TSA data ($\Delta T/T$ is linearly proportional to the stress) obtained when crack reaches the tri-material junction. Figure 15.2a was obtained from C1. Large

$\Delta T/T$ values are produced in the face sheet, on the right hand side, ahead of the crack stopper. In the results obtained from C3 (see Fig. 15.2b), the $\Delta T/T$ value ahead of the crack stopper tip is much smaller. The localized stress concentration ahead of the crack stopper tip can be an explanation for the crack propagation behaviour observed in C1.

## 15.3 Numerical Fracture Mechanics Study

A numerical fracture mechanics study was conducted using the commercial FE package ANSYS 15.0. The two scenarios of crack deflection and crack propagating along the horizontal interface (i.e. no crack deflection) at the tri-material junction were modelled for each peel stopper configuration. 8-node 2D plane stress elements (PLANE 183) with an average element size of 0.5 mm were used. Near the crack tip, the number of elements ranged from 36 to 144 with element sizes of 5–10 μm. Figure 15.3 shows the FE meshes corresponding to the three peel stopper configurations and the geometry of the sandwich beam model. The detailed models of different crack path scenarios around the tri-material junction for each configuration are shown in the images (a)–(f) in Fig. 15.3.

In particular, the case of crack deflection for the crack passing the tri-material junction was analysed. The predicted energy release rate and the mode-mixity values were compared to the experimentally observed crack paths. The paths

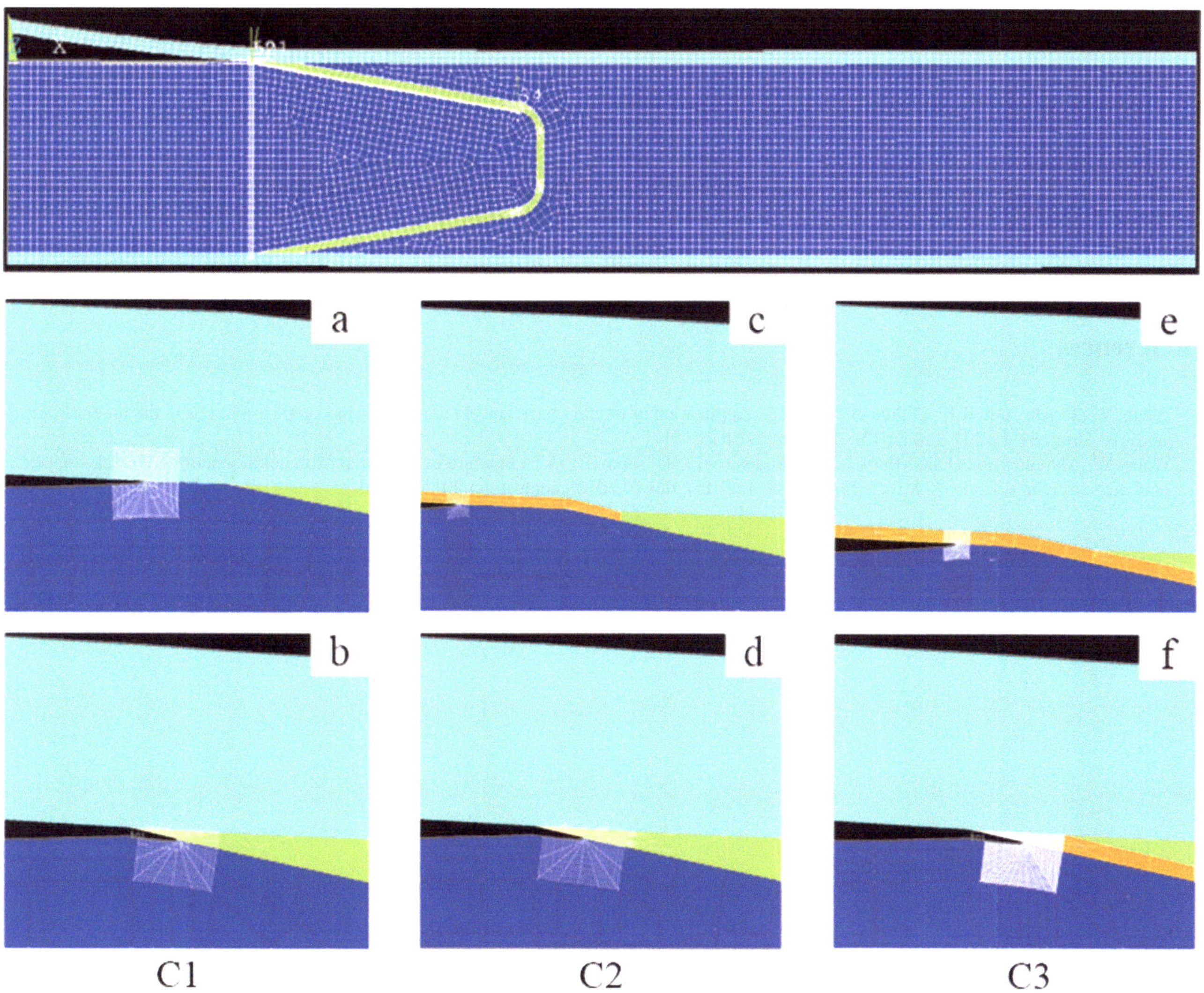


**Fig. 15.3** FE models of the sandwich specimens corresponding to different crack path scenarios near the tri-material junction for C1 (**a** and **b**), C2 (**c** and **d**) and C3 (**e** and **f**)

consist of the "straight path" for the crack penetrating the peel stopper, and the "deflected path" for the crack deflecting at the peel stopper. The ability to achieve crack deflection wrt. the peel stopper angle $\theta$ was also investigated. The peel stopper angle, $\theta$ was varied from 5° to 30° with steps of 5°, and with a kinked crack length of 0.5 mm away from the corner as shown in Fig. 19. The 10° angle represents the tested configuration. The numerical study has shown that by increasing the peel stopper angle, crack deflection at the tri-material junction becomes increasingly more difficult. In the study the 10° angle was the one tested experimentally, since it represents the more practical solution.

## 15.4 Summary and Conclusions

It has been shown that the change of the crack stopper configuration results in different local effects at the tri-material junction which affect the crack propagation path. A numerical fracture mechanics study using finite element modelling was conducted to analyse and interpret the data so that a detailed fracture mechanics analysis could be used to explain the experimental observations. For the tested 10° angle, the energy release rate for the two possible crack paths, crack propagating straight and crack deflecting at the peel stopper were used to derive results on the crack deflection ability of each configuration. It was shown that if the ratio of the energy release rates for the two crack paths is equal (or near equal) to the ratio of the interface fracture toughnesses of the two crack paths, then crack deflection at the peel stopper is unlikely to occur as observed for C1. To ensure crack deflection it was found that the interface fracture toughness of the straight path must be large compared to the interface fracture toughness of the deflected crack path. By placing/embedding fibres in front of the peel stopper tip in C2 and C3 the desired behavior was achieved and the crack is deflected every time, as was confirmed in the experiments.

The work presented is based on recently published work [1, 2].

**Acknowledgements** The work presented was co-sponsored by the University of Southampton and the Danish Council for Independent Research | Technology and Production Sciences (FTP), under the research project 'Enhanced Performance of Sandwich Structures by Improved Damage Tolerance' ('SANTOL'). The financial support received is gratefully acknowledged. The foam material supported by DIAB AB Sweden is highly appreciated.

## References

1. Wang, W., Fruehmann, R.K., Dulieu-Barton, J.M.: Application of digital image correlation to address complex motions in thermoelastic stress analysis. Strain **51**(5), 405–418 (2015). doi:10.1111/str.12151
2. Wang, W., Martakos, G., Dulieu-Barton, J.M., Andreasen, J.H., Thomsen, O.T.: Fracture behaviour at tri-material junctions of crack stoppers in sandwich structures. Compos. Struct. **133**, 818–833 (2015). doi:10.1016/j.compstruct.2015.07.060

# Chapter 16
# Probabilistic Improvement of Crack Propagation Monitoring by Using Acoustic Emission

**Malick Diakhate, Emilio Bastidas-Arteaga, Rostand Moutou Pitti, and Franck Schoefs**

**Abstract** In this work, acoustic emission (AE) is used as a measurement technique to detect and locate the crack tip and, to monitor its propagation in a wooden specimen subjected to mechanical stresses. Tensile tests were performed on DCB specimens to generate mode I cracking. Under these stresses, the material response results in a release of energy in the form of transient elastic waves that were recorded by AE sensors. By means of the AE technique, the monitoring of material damage lies in the ability to identify the most relevant descriptors of cracking mechanisms. The latter are identified by clustering the AE data. A K-means++ algorithm was used, and two AE features—peak frequency and number of counts—represent adequately the AE events clustering. This unsupervised classification allows the AE events that were generated by crack tip growth during the test to be identified. There are many parameters that can affect the accuracy of AE monitoring such as noise signals, geometry, wood species, etc. Consequently, in this study, probabilistic approaches (Probability of Detection) were used to both characterize uncertainties and improve AE experimental protocol.

**Keywords** Acoustic emission • Crack tip monitoring • Cluster analysis • Probability of detection • Wood material

## 16.1 Introduction

Due to the manufacturing process of building materials some defects are presents in civil engineering and mechanical structures. These manufacturing defects could propagate under environmental or mechanical loadings, and could induce the total collapse of the structure. It is a well known fact that timber elements exhibit micro-cracks [1] which can propagate under fatigue, overload, or creep loading, and can cause failure of the structure [2]. In addition, wood is a hygroscopic material whose mechanical behavior is very sensitive to climatic changes such as temperature and moisture content variations [3]. For example, drying process accelerates the crack growth, whereas humidification process lead to a delay in crack propagation [4].

In order to bring some responses to these scientific problems, the ANR JCJC2013 CLIMBOIS project [5] is dealing with the effects of climatic and mechanical variations on the durability of timber structures. As part of this research project devoted to both an identification of failure mechanisms in wood material and an evaluation of the crack length evolution during tests, the study carried out herein included both experimental tests in constant and variable environmental conditions and statistical and probabilistic analysis of the acoustic emission (AE) results.

Recent years witnessed a dynamic development of the acoustic emission method. It finds a wider and wider application in many industrial fields, mainly to monitor the structural health of the structure components [6–8]. In fracture mechanics, some authors have applied the AE technique in order to identify the process zone [9] or to monitor the fracture parameters in wood parameters [10]. Acoustic Emission has also been applied for tracing fracture intensity in lime wood due to climatic variations [11] and in dried specimen [12]. But the specimen that was used didn't provide a stable crack growth during crack tip propagation.

M. Diakhate (✉)
Université Bretagne Occidentale, FRE CNRS 3744, IRDL, 43 Quai de Léon, 29600 Morlaix, France
e-mail: malick.diakhate@univ-brest.fr

E. Bastidas-Arteaga • F. Schoefs
LUNAM, Institute for Research in Civil and Mechanical Engineering GeM, Université de Nantes, CNRS UMR 6183/FR 3743, 2 Rue de la Houssinière BP 92208, F-44332 Cedex 3 Nantes, France

R.M. Pitti
Institut Pascal, Université Clermont Auvergne, Université Blaise Pascal, BP 20206, 63000 Clermont-Ferrand, France

Institut Pascal, CNRS, UMR 6602, 63171 Aubière, France



A.T. Zehnder et al. (eds.), *Fracture, Fatigue, Failure and Damage Evolution, Volume 8*,
Conference Proceedings of the Society for Experimental Mechanics Series, DOI 10.1007/978-3-319-42195-7_16

This paper presents the results from the laboratory tests that were devoted to the monitoring of crack tip propagation within wood material by using a stable growth specimen. By means of the AE technique, the monitoring of material damage lies in the ability to identify the most relevant descriptors of cracking mechanisms. The latter are identified by clustering the AE data. A K-means++ algorithm [8, 13] was used, and two AE features—peak frequency and number of counts—represent adequately the AE events clustering. This unsupervised classification allows the AE events that were generated by crack tip growth during the test to be identified.

There are many parameters that can affect the accuracy of AE monitoring such as noise signals, geometry, wood species, etc. Consequently, in this study, probabilistic approaches (Probability of Detection) were used to both characterize uncertainties and improve AE experimental protocol [14].

## 16.2 Materials and Testing Conditions

This section is devoted to a description of the laboratory tests that were conducted to investigate the effectiveness of using the AE technique to monitor the crack tip propagation within wood material. The experimental conditions (wood species, specimens, mechanical loading conditions, etc.) are described. Both the AE equipment specifications and the sample preparation are presented. Figure 16.1 provides a general overview of the experimental set-up.

### *16.2.1 Wood Materials, DCB Specimen, and Testing Conditions*

In this study, the two species used in Europe (Douglas and White Fir) were tested to identify their failure mechanisms under mechanical loadings and constant moisture content (around 9 %).

The wood sample geometry is a Double Cantilever Beam (DCB) in order to provide the stability of the energy release rate during the crack growth process under opening mode solicitation [3]. Several samples of each species have been machined in a Radial-Longitudinal configuration. A pre-crack of 50 mm in length (Fig. 16.1b) was performed along the grain direction with a band saw (3 mm thick). The symmetric of both the geometry and the loading allow assuming an open mode configuration according to a particular choice of grain alignment with the crack.

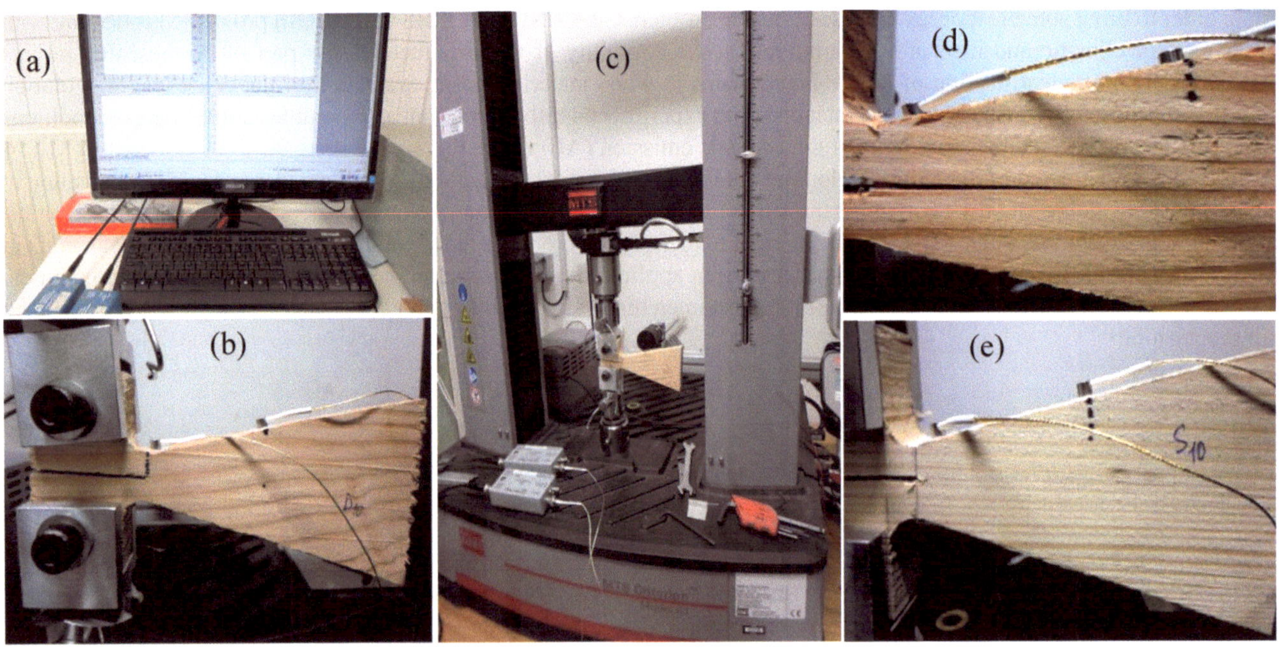


**Fig. 16.1** Material, DCB specimen, and experimental setup: (**a**) AE equipment, (**b**) DCB specimen (Douglas fir) with AE sensors, (**c**) testing machine, (**d**) crack propagation during test, (**e**) DCB specimen (White fir) with AE sensors

An electromechanical MTS® press (Fig. 16.1c) was used to apply the loading. The machine is equipped with a cell force (±500 N) and a displacement sensor. Two linking parts made of aluminium were manufactured (Fig. 16.1b), and allow connecting the specimen to both the jack and the cell force of the testing machine. The tests were performed at a constant displacement rate of 0.5 mm/min. This displacement control allows forcing stable crack growth during the test. During the test, both resulting force and applied displacement values were recorded with a data acquisition system.

### *16.2.2 Acoustic Emission Method*

Under mechanical loading, the wood cracks generate transient elastic waves. The latter are referred to as AE waveforms. Four-channel AE system designed by MISTRAS Group (Fig. 16.1a) were used to record the AE waveforms that were generated during the laboratory tests. Since the crack path within the material is expected to follow the grain direction, two AE channels were enough to both record and perform a linear localization of the AE sources (referred to as events). Thus, two piezoelectric sensors (Fig. 16.1d, e), with a frequency range 200–750 kHz (450-kHz resonant frequency) were connected to two preamplifiers (IL40S model, gain set at 40 dB). These preamplifiers were connected to the data acquisition card of the AE system. The AE sensors are coupled to the specimen with a double-face adhesive tape. The data acquisition threshold was set at 35 dB. The AE waveforms were sampled at 40 MHz, and the wave propagation velocity within the wood material was set at 4000 m/s. This velocity allowed 1D-localization of the AE sources. The effectiveness of this 1D-localization was evaluated in terms of Probability of Detection [14].

### *16.2.3 Repetitive Pencil Breaking Tests and Assessment of Probability of Detection (PoD)*

The experimental procedure aims at both quantifying and improving the accuracy of the estimated locations following the generation of AE waves at the known positions at the fracture surface of the wood specimen. These known positions simulate the crack tip positions (AE sources). AE waves were generated at various pre-defined known positions by the normalized Hsu-Nielsen sources (2H—0.5 mm pencil lead) (Fig. 16.2a) according to the EN 1330-9 standard [15, 16]. The pencil lead breaking test allows evaluating the AE wave propagation velocities within the wood specimen. The AE wave propagation velocities strongly depend on propagation direction (along or perpendicular to the grain) within the orthotropic material.

To evaluate the reliability of AE measurements, we compared the known crack tip with the AE estimated positions (given by the location algorithm program). Several tests were repeated at the same known position to quantify measurement and material uncertainties in terms of the probability of detection (PoD).

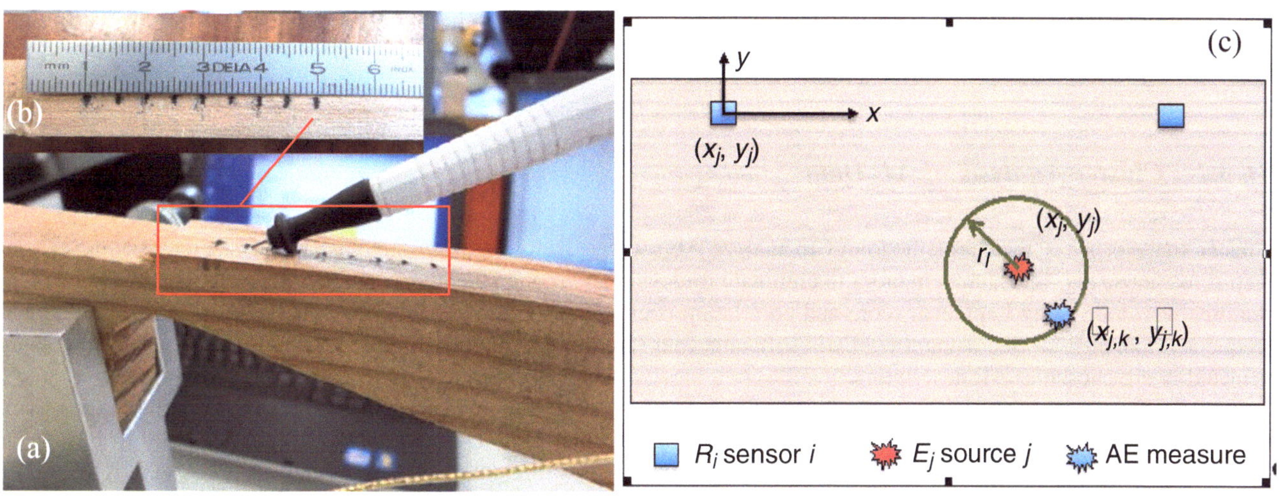


**Fig. 16.2** (**a**) Half-DCB specimen (Douglas fir), (**b**) position of Hsu-Nielsen sources, (**c**) problem parameterization

The pencil lead breaking tests were performed on a half-DCB specimen (Fig. 16.2a). The seven positions shown in Fig. 16.2b were selected on the middle axis of the fracture surface of the half-DCB specimen (1, 1.5, 2, 2.5, 3, 3.5 and 4 cm) to simulate the evolution of crack tip positions during crack propagation. At each known position (Hsu-Nielsen sources), the pencil lead breaking was performed several times to record enough AE signals for probabilistic analysis.

Each pencil lead breaking is denoted as an event d. Figure 16.2c shows the problem parameterization that was considered in the present study. The events d were positioned at $j = 7$ known points $(x_j, y_j)$. Thirty measurements ($n = 30$) were realized at each j location. Each measure provided a kth AE coordinate $\left(\tilde{x}_{j,k}, \tilde{y}_{j,k}\right)$.

Among all uncertainty sources, the event d is considered as deterministic in intensity and position (by considering a positioning error less than 0.5 mm). The events d were repeated to quantity the measurement error that integrates the uncertainties related to the sensor and the source position for a fixed AE configuration $AE_{con}$ (distance between sensors, post-treatment algorithm, velocity along the grain, etc.). Since the real position was known, it was possible to quantify the bias and the uncertainty for each position or configuration. This information is useful to determine correction functions and AE configurations that minimize the measurement error.

Considering that the material is heterogeneous, material uncertainties were integrated to each measure point. In addition, if tests are considered as equally probable and sufficiently representative of real situations, the PoD of a position j is then written with respect to the membership of the measured position of the source to a circle of radius $r_l$ centered on the exact coordinates of the source (Fig. 16.2c).

$$PoD_{j,AE_{con}}(d, r_l) = P\left(\sqrt{\left(x_j - \tilde{x}_{j,k}\right)^2 + \left(y_j - \tilde{y}_{j,k}\right)^2} \leq r_l \middle| d\right) \tag{16.1}$$

The value of $r_l$ can be considered as the required detection threshold for the localization of the crack tip. This study will focus on a fixed AE configuration, $AE_{con}$. The current configuration of the sensors (Fig. 16.1) allows only estimating the position of the source in one dimension. Consequently, we focus on the detection of the position in the x-direction. In such a case, Eq. (16.1) becomes:

$$PoD_{j,AE_{con}}(d, r_l) = P\left(\left|x_j - \tilde{x}_{j,k}\right| \leq r_l \middle| d\right) \tag{16.2}$$

In this study the PoD was directly evaluated from the n repetitions of the AE measurement at each j position.

## 16.3 Results Analysis and Discussion

This section is devoted to a description of the statistical tools that were used to both identify the failure mechanisms and evaluate the crack tip length during mechanical tests. The recorded AE waveforms were analyzed by means of the AEwin software developed by MISTRAS Group. Nine AE features were selected for the data clustering process. Figure 16.3 provides a general overview of the results from the cluster analysis of AE data.

### *16.3.1 Cluster Analysis of AE Data*

Cluster analysis is a statistical methodology to analyze AE sources. Before performing cluster analysis, the first and crucial step is to choose the AE features in order to eliminate irrelevant and redundant signals [8, 13]. The selection of the relevant features was done using Laplacian score and correlation coefficients. Thus, from the nine AE features (time domain and frequency domain) selected using the AEwin software, four (amplitude, number of counts, peak frequency, and frequency centroid) were chosen as AE descriptors.

Among cluster algorithms, the most frequently used methods are K-means, self-organized map combination with k-means and fuzzy-c means algorithm. K-means is the simplest and most effective method for AE signal clustering. To perform cluster analysis of the AE events recorded during the tests, Principal Component Analysis (PCA) and K-means++ algorithm were used to gather similar AE events into clusters. The cluster analysis was performed using the Statistics toolbox and functions in the SOM toolbox in Matlab™ software [8, 13]. The optimal number of clusters is chosen based on

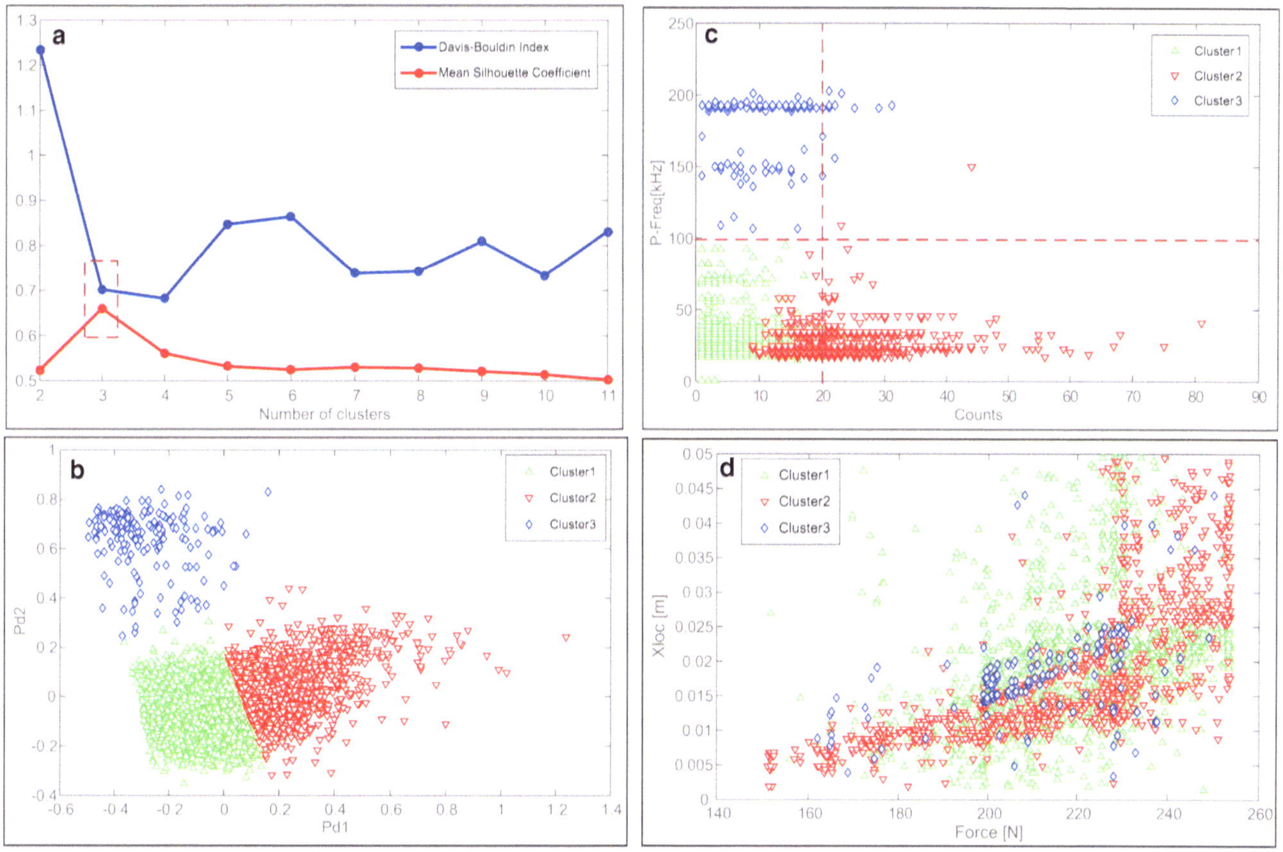


**Fig. 16.3** Cluster analysis of AE data: (**a**) optimization of the number of clusters, (**b**) PCA visualization of K-means++ clustering, (**c**) clustering of AE events using two AE features, (**d**) linear location of clusters during mode I test

the values of both the Davis–Bouldin index (DB) and the Silhouette Coefficient (SC). The optimal number of clusters leads to a highest value of SC and a lowest value of DB. With these values, the clusters are dense and well separated, which corresponds to the standard concept of a cluster [8].

The cluster analysis presented shows that the optimal number of clusters is three (Fig. 16.3a). Figure 16.3b presents the projection of the three clusters of AE events onto a two-dimensional plot by two principal components. The two AE features, peak-frequency (P-Freq) and number of counts (Counts) are the most important AE parameters in the chosen set of four, as evidenced by the good separation of the clusters points in the space of these two parameters (Fig. 16.3c). AE events of Cluster#1 exhibit a peak-frequency lower than 100 kHz, and a number of counts lower than 20. AE events of Cluster#2 exhibit a number of counts greater than 20, and a peak frequency lower than 100 kHz. AE events of Cluster#3 exhibit a peak frequency greater than 100 kHz, and a number of counts lower than 20. Figure 16.3d shows the linear localization of the AE events of the three clusters during the mechanical test.

Based on Fig. 16.3d, the findings are as follows: AE events of Cluster#1 are linked to wood micro-cracks; AE events of Cluster#2 are linked to crack tip propagation. At the current stage of the cluster analysis, and on the basis of observation of fracture surfaces of specimens, AE events of Cluster#3 could be linked to wood-fiber breaks.

### *16.3.2 Crack Tip Propagation*

Figure 16.4 presents different graphs to illustrate the crack tip propagation during the test. By only selecting AE events of Cluster#2 (specific values of the two AE features: peak-frequency and number of counts), the crack tip position (X-Loc) can be plotted versus the corresponding force (Fig. 16.4a). Due to the geometry shape of the specimen, the known positions (X-Real at several positions on the fracture surface) are different from those estimated by the AE system (X-Loc). Hence,

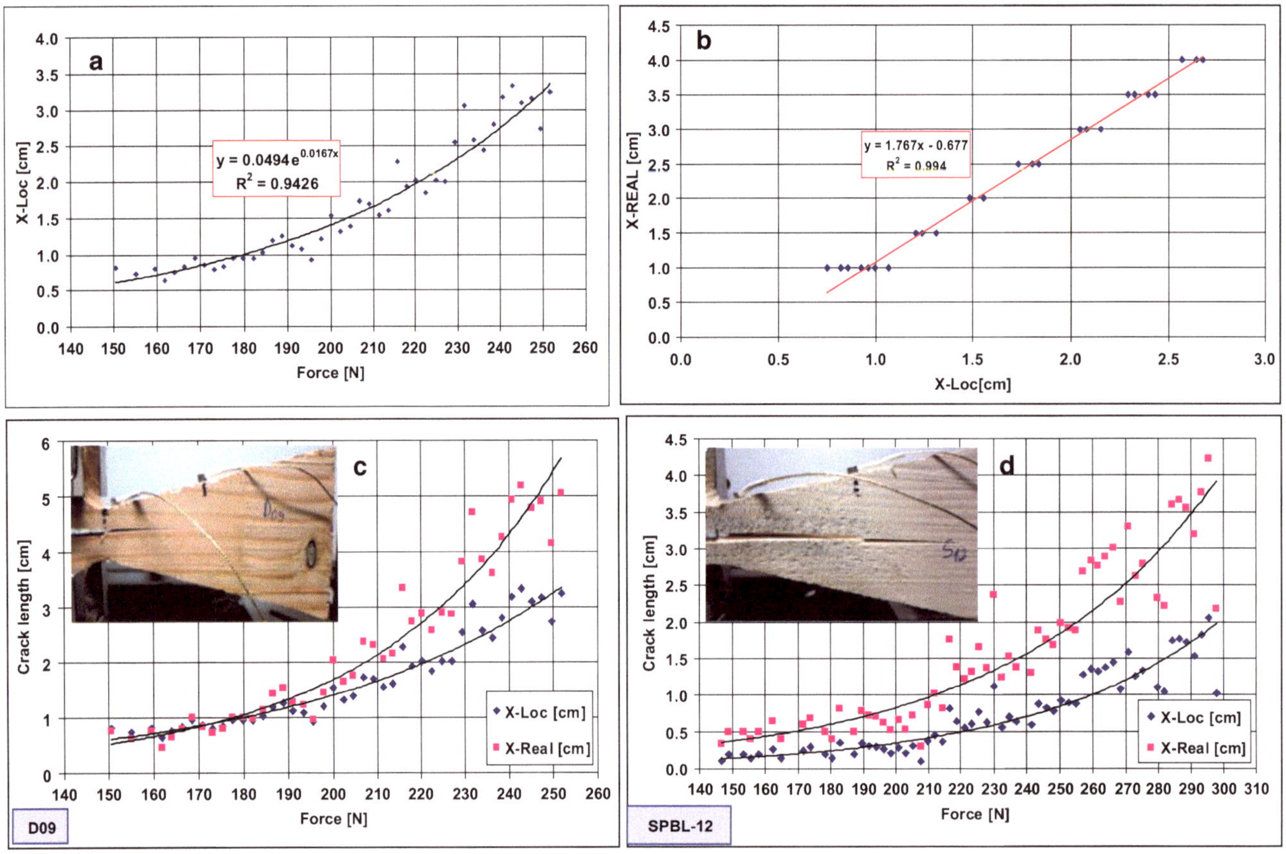


**Fig. 16.4** AE events of Cluster #2 (crack tip propagation during test): (**a**) AE results, (**b**) correlation law between pencil breaking known positions (X-Rael) and AE events locations (X-Loc), (**c**, **d**) crack tip propagation respectively for Douglas fir wood and White fir wood

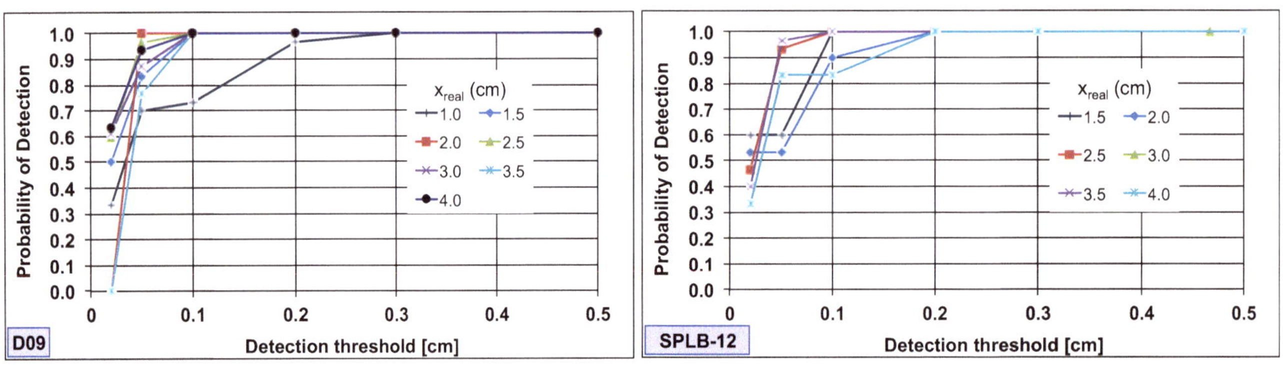


**Fig. 16.5** Probability of detection for crack tip propagation: Douglas fir wood (D09) and White fir wood (SPLB-12)

pencil lead breaking protocol was applied to provide a correlation law that links the X-Real values to X-Loc ones (Fig. 16.4b). This correlation law allows plotting the X-Real (crack tip real position) versus the corresponding force (Fig. 16.4c, d).

### *16.3.3 Probability of Detection*

The repetitive tests described in §2.3 were also used to evaluate the PoD as a function of detection thresholds (Fig. 16.5). This detection threshold was defined as an indicator of the precision of the AE measures. Figure 16.5 presents the PoD for two specimens (Douglas fir wood and White fir wood) and various detection thresholds and positions ($X_{real}$). It is observed for both specimens and all positions that the PoD is larger than 50 % for a detection threshold of $r_l = 0.1$ cm. It becomes larger than 90 % for $r_l$ greater than 0.2 cm. It is also noted that the PoD is not influenced by the position of the acoustic source $X_{real}$. These results indicate that AE results corrected by repetitive tests are useful for monitoring crack propagation particularly for applications when the required precision of detection is lower than 2 mm.

## 16.4 Conclusion

Throughout their service life, timber structures will be subjected to both mechanical and environmental loadings. The lack of crack tip monitoring within wood material when the structure is subjected to variable environmental conditions could lead to the collapse of the structure, in particular due to the fact that wood mechanical behavior is very sensitive to climatic changes such as temperature and moisture content variations. In this paper, the monitoring of the crack tip propagation within wood material is investigated using the Acoustic Emission method. DCB samples (Douglas and White Fir) were tested on an electromechanical machine and under constant moisture content (around 4 %). Tensile tests were performed on these DCB samples to generate mode I cracking. Under these stresses, the material response results in a release of energy in the form of transient elastic waves that were recorded by AE sensors.

AE events recorded during the tensile tests on wood samples were subjected to cluster analysis, which results in identification of clusters of AE events. This allows the failure mechanism which caused an AE event to be identified. Two AE features—peak frequency and number of counts—represent adequately the AE events clustering. This unsupervised classification allows the AE events that were generated by crack tip propagation to be identified.

In this study, the findings are as follows. AE events of Cluster#1 exhibit a peak-frequency lower than 100 kHz, and a number of counts lower than 20. These AE events are linked to wood micro-cracks. AE events of Cluster#2 exhibit a number of counts greater than 20, and a peak frequency lower than 100 kHz. These AE events are linked to crack tip propagation. AE events of Cluster#3 exhibit a peak frequency greater than 100 kHz, and a number of counts lower than 20. At the current stage of the cluster analysis, and on the basis of observation of fracture surfaces of specimens, AE events of Cluster#3 could be linked to wood-fiber breaks. AE events of Cluster#2 have allowed the crack tip propagation to be monitored. For all samples, the acoustic signature of the crack tip propagation within the wood material was characterized by a peak frequency lower than 100 kHz and a number of counts greater than 20. Since the acoustic signature of the crack tip propagation has been identified, the effect of different moisture contents on crack tip propagation within wood material can now be investigated. Effects of moisture content on acoustic signature will be also evaluated.

The effectiveness of AE measurements (events 1D-localization) has been evaluated in terms of Probability of Detection. Repetitive pencil breaking tests were performed on each half-DCB sample (after DCB sample failure). This experimental procedure aims at both quantifying and improving the accuracy of the estimated locations following the generation of AE waves at the known positions at the fracture surface of the wood specimen. The repetitive tests were also used to evaluate the PoD as a function of detection thresholds. The PoD has been evaluated for different values of detection threshold. Results showed that the PoD is larger than 50 % for a detection threshold of $r_l = 0.1$ cm. The PoD value becomes larger than 90 % for $r_l$ greater than 0.2 cm. These results indicate that AE results corrected by repetitive tests are useful for monitoring crack propagation particularly for applications when the required precision of detection is lower than 0.2 cm.

**Acknowledgments** The authors wish to strongly thank the National Agency of Research (ANR) for its financial support of this work through the project CLIMBOIS N°ANR-13-JS09-0003-01 labeled by ViaMeca. The authors also thank Li Li (Donghua University, China) and Stephan Lomov (KU Leuven, Belgium) for the cluster analysis tool.

## References

1. Riahi, H., Moutou Pitti, R., Dubois, F., Fournely, E.: On numerical evaluation of mixed mode crack propagation coupling mechanical and thermal loads in wood material. In: Carroll, J., Daly, S. (eds.) Fracture, Fatigue, Failure, and Damage Evolution, vol. 5. pp. 21–26. Springer, New York (2015)
2. Moutou Pitti, R., Hamdi, S.E., Dubois, F., Riahi, H., Angellier, N.: Numerical fracture analysis under temperature variation by energetic method. In: Beese, A.M., Zehnder, A.T., Xia, S. (eds.) Fracture, Fatigue, Failure and Damage Evolution, vol. 8, pp. 243–251. Springer, New York (2016)
3. Angellier, N., Moutou Pitti, R., Dubois, F.: Crack analysis of wood under climate variations. In: Beese, A.M., Zehnder, A.T., Xia, S. (eds.) Fracture, Fatigue, Failure and Damage Evolution, vol. 8, pp. 235–242. Springer, New York (2016)

4. Saifouni, O., Moutou Pitti, R., Destrebecq, J.F.: Determination of stresses in drying wood by means of a viscoelastic relaxation model. In: Antoun, B., Qi, H.J., Hall, R., Tandon, G.P., Lu, H., Lu, C. (eds.) Challenges in Mechanics of Time-Dependent Materials and Processes in Conventional and Multifunctional Materials, vol. 2, pp. 29–36. Springer, New York (2013)
5. Moutou Pitti, R., Diakhaté, M., Bastidas-Arteaga, E., Aoues, Y., Angellier, N., Riahi, H., Chateauneuf, A., Dubois, F.: ANR JCJC-2013 Project CLIMBOIS: effects of climatic and mechanical variations on the durability of timber structures: use of acoustic emission tool to evaluate wood mechanical behaviour. In: COST Action FP1302 Wood Music, Opening Conference, Paris, 26–28 February 2014
6. Diakhaté, M., Larcher, N., Takarli, M., Angellier, N., Petit, C.: Acoustic techniques for fatigue cracking mechanisms characterization in hot mix asphalt (HMA). In: 7th RILEM International Conference on Cracking in Pavements, Springer, vol. 4, pp. 771–781 (2012)
7. Hamdi, S.E., Le Duff, A., Simon, L., Plantier, G., Sourice, A., Feuilloy, M.: Acoustic emission pattern recognition approach based on Hilbert-Huang transform for structural health monitoring in polymer-composite materials. Appl. Acoust. **74/5**, 746–757 (2013)
8. Li, L., Lomov, S., Yan, X.: Correlation of acoustic emission with optically observed damage in a glass/epoxy woven laminate under tensile loading. Compos. Struct. **123**, 45–53 (2015)
9. Lamy, F., Takarli, M., Angellier, N., Dubois, F., Pop, O.: Acoustic emission technique for fracture analysis in wood materials. Int. J. Fract. **192**(1), 57–70 (2015)
10. Reiterer, A., Stanzl-Tschegg, S.E., Tschegg, E.K.: Mode I fracture and acoustic emission of softwood and hardwood. Wood Sci. Technol. **34**(5), 417–430 (2000)
11. Jakieła, S., Bratasz, L., Kozłowski, R.: Acoustic emission for tracing fracture intensity in lime wood due to climatic variations. Wood Sci. Technol. **42**(4), 269–279 (2008)
12. Kowalski, S.J., Molinski, W., Musielak, G.: The identification of fracture in dried wood based on theoretical modelling and acoustic emission. Wood Sci. Technol. **38**(1), 35–52 (2004)
13. Li, L., Lomov, S., Yan, X., Carvelli, V.: Cluster analysis of acoustic emission signals for 2D and 3D woven glass/epoxy composites. Compos. Struct. **116**, 286–299 (2014)
14. Schoefs, F., Boéro, J., Clément, A., Capra, B.: The αδ method for modelling expert judgement and combination of non-destructive testing tools in risk-based inspection context: application to marine structures. Struct. Infrastruct. Eng. **29**, 531–543 (2012)
15. Hsu, N.N., Breckenridge, F.R.: Characterization and calibration of acoustic emission sensors. Mater. Eval. **39**, 60–68 (1981)
16. Sause, M.G.R.: Investigation of pencil lead breaks as acoustic emission sources. J. Acoust. Emiss. **29**, 184–196 (2011)

# Chapter 17
# Failure Detection of Temporary Structures with Digital Image Correlation for Construction Safety Applications

**Shaowen Xu**

**Abstract** Temporary structures refer systems and assemblies without a permanent foundation, which will be removed after a certain period of time. In the past decades there have been numerous significant collapses of temporary structures in United States. To ensure structural stability and the safety, there is increasingly demand of developing an industry wide standard code, technology and procedure of temporary structure design, erecting, maintenance, monitoring and evaluation for practicing engineers and contractors. In this study, Digital Image Correlation was utilized for structure monitoring and failure detection of temporary structure during construction or loading period. The experimental results demonstrated the displacement measurement at the structural nodes could be used to determine the safety stage of the structure, predict the structural failure time and location. The aim of this research is to advance this state of the art measurement technology and facilitate a real-time automated monitoring and detection system for of structural failures of temporary structures.

**Keywords** Structure monitoring • Failure detection • Digital image correlation • Construction safety • Temporary structure

## 17.1 Introduction

Temporary structures are low cost assemblies and systems, without a permanent foundation, used for temporary supporting or bracing of permanent work during construction, or structures built for temporary use, such as temporary bracing, scaffolding, temporary or emergency shelters, temporary formwork and performance stage. In the past decades there have been significant numbers of collapses of temporary structures in the United States. Those accidents resulted in many deaths and injuries, as well as, huge financial lost [1–5]. Investigations indicated that many of those accidents related to improper design, erection and monitoring, as well as, unpredicted loading conditions. To ensure structural stability and safety, there is increasingly demand of developing an industry wide standard code, and technology of temporary structure design, erecting, monitoring and evaluation for practicing engineers and contractors.

The performance status of a temporary structure could characterize in terms of *stable*, *transition*, and *unstable stage*. In *stable stage* the structure is considered reliable and safety. The structure would tend to recover back to its equilibrium state after any perturbation. After passing the "*Instable Threshold*" into *transition stage*, the structure will tend to deform away from its equilibrium state during loading process. When load further increasing, the structure will become unstable. Without any additional support, the structure would quickly switch into *unstable stage*, deform rapidly and collapse when it experiences any perturbation of load and external introduced displacement. The critical point between the *transition* and *unstable stages* is referred as *Failure Point*.

Structural failures are the result of stress in the structure reaching a critical level due to combinations of causes, such as, accumulated loading, local structural damage, poorly located slabs and unpredicted dynamic loading. These factors are not well recognized and difficultly determined in construction sites for temporary structures, since the limitations of the available technologies, and safety standards have not been well developed yet.

Digital Image Correlation (DIC) technology is a non-contact optical method for full field measurements by tracking the changes in images [6–9]. It could be applied in either in macro-scale or micro-scale [10–12] to provide accurate full deformation and strain fields on static or dynamic loading condition. It could be also integrated with numerical simulation for the applications in extreme loading condition [13]. In this study, the DIC was utilized to monitor structural deformation and strain during construction or loading period. The aim of this research was expecting to develop a new methodology and facilitate a real-time automated system for structure evaluation, monitoring and failure detection, which could prevent

S. Xu (✉)
Department of Mechanical Engineering, Georgia Southern University, Statesboro, GA, USA
e-mail: shaowenxu@georgiasouthern.edu



A.T. Zehnder et al. (eds.), *Fracture, Fatigue, Failure and Damage Evolution, Volume 8*,
Conference Proceedings of the Society for Experimental Mechanics Series, DOI 10.1007/978-3-319-42195-7_17

accidents in construction sites and reduce fatalities and injuries. Such method and system could be able to capture critical points of performance status of a temporary structure, and give warning signals to construction workers before a potential failure occurs. In addition to failure prediction, the proposed method could also support the efforts of developing more efficient regulations, and standards to guide construction and management of civil infrastructure in long term.

## 17.2 Experiment Set Up

Experiments were designed and conducted using small-scale temporary structures as shown in Fig. 17.1. The frame of the structure was built using pieces of metal drywall corner bead, hex bolts and nuts. Mimicking a real-world situation, the tested structure was not made to be perfectly conditions in terms of material properties and symmetry, etc. Load was applied by adding water in the plastic bins on the structure slowly to mimic the loading of fresh concrete. Presented in this paper is one of such experiments. In this study, a bin was set on the right front top of the structure. Random spot pattern was printed on the front surface of the structure for digital image processing.

In the experiment, two cameras were used to monitor the structural response of the structure during loading. With the DIC system, (1) detailed deformation histories of the structure were determined, (2) the precise location of the initial failure was identified, (3) the instable threshold and Failure point were predicted. It is noted that presented here are only the results of 2D DIC analysis, which were processed based on the images of one camera. The results of 3D DIC will be presented in other technical paper.

## 17.3 Results and Discussion

Shown in Fig. 17.2 were the DIC displacement (Fig. 17.2a, b) and strain (Fig. 17.2c, d) field contours of the structure at selected time steps (110, 150, and 160). At the step 110, the displacement and strain fields were relative uniform, the maximum magnitude of displacement were 0.113 mm (Δu) in horizontal 0.219 mm (Δv) in vertical; the maximum magnitude of Lagrange Strain were 0.0035 ($e_{xx}$) in horizontal 0.0125 ($e_{yy}$) in vertical. At the step 150, there were sudden large increments of all these quantities, 0.600 mm for Δu, 0.268 mm for Δv, 0.11 for $e_{xx}$ 0.010 for $e_{yy}$. The displacement and strain fields became more non-uniform. Stating from this point, the top of the structure began to move faster towards the left. When load increased and reached the step 160, it was observed that larger horizontal displacement was found at the left side of the top-edge, and the structure had shifted to the left 2.70 mm. Larger vertical displacement (0.43 mm) was also found on the middle of the right edge. When examining the strain fields, larger vertical strain (0.030) was found developing fast at the right bottom edge. It could be seen that the displacement of top-front-edge was exhibiting an increasing displacement towards the left; right-front-edge had increasing displacement towards ground. The initial failure location was at near the node 3.

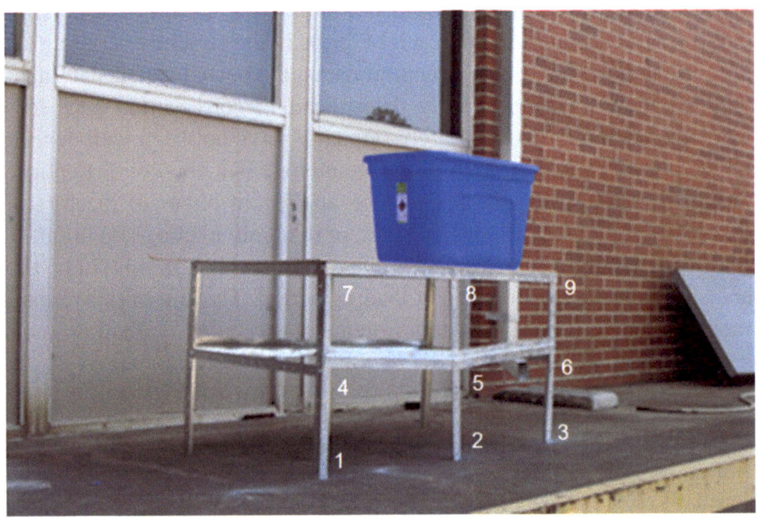


**Fig. 17.1** Temporary structure used in the experiment

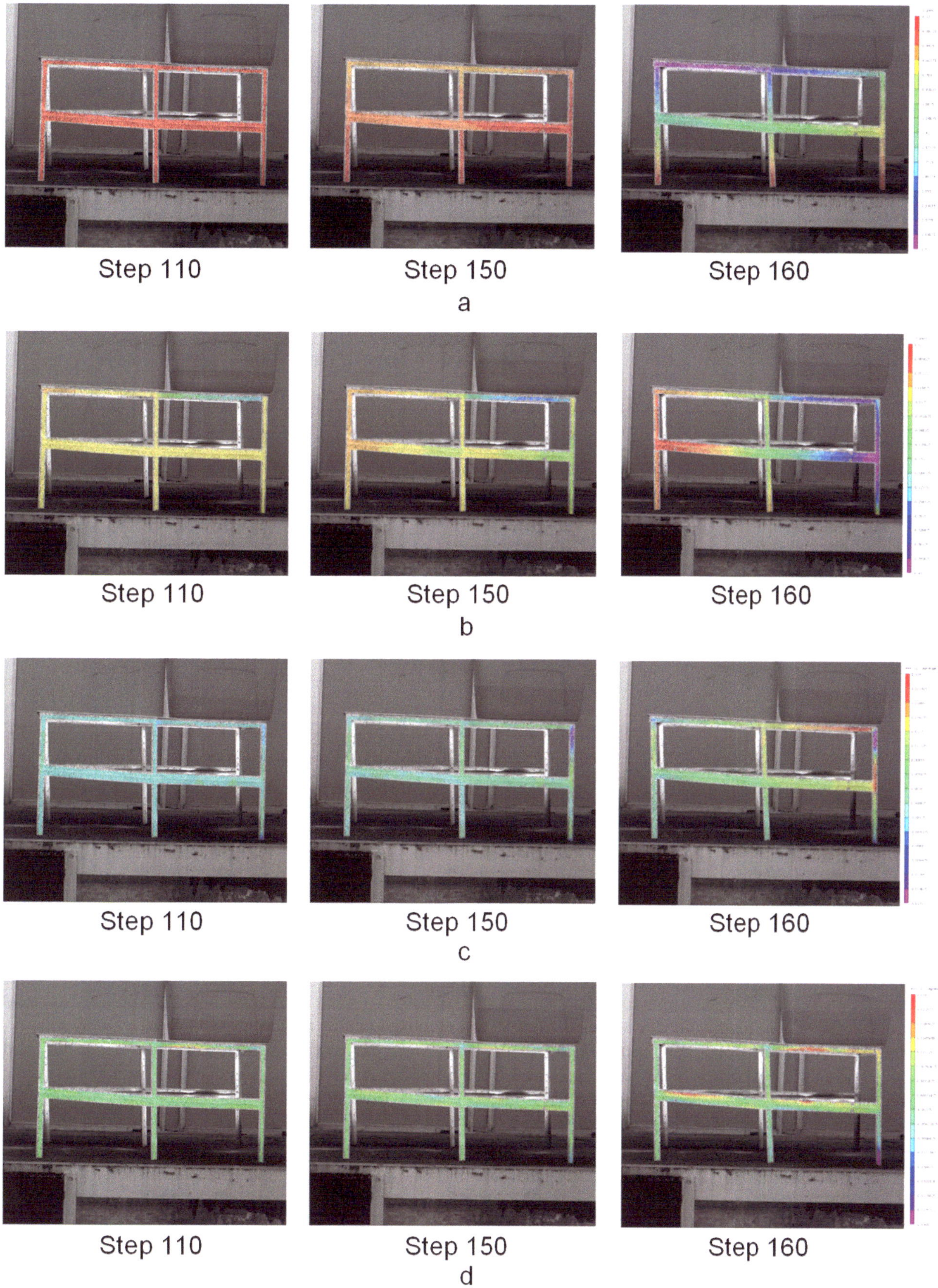


**Fig. 17.2** DIC color contours of displacement (**a**) in horizontal and (**b**) vertical direction and Lagrange strain in (**c**) horizontal and (**d**) vertical direction

Fig. 17.3 Horizontal displacement history at structural nodes

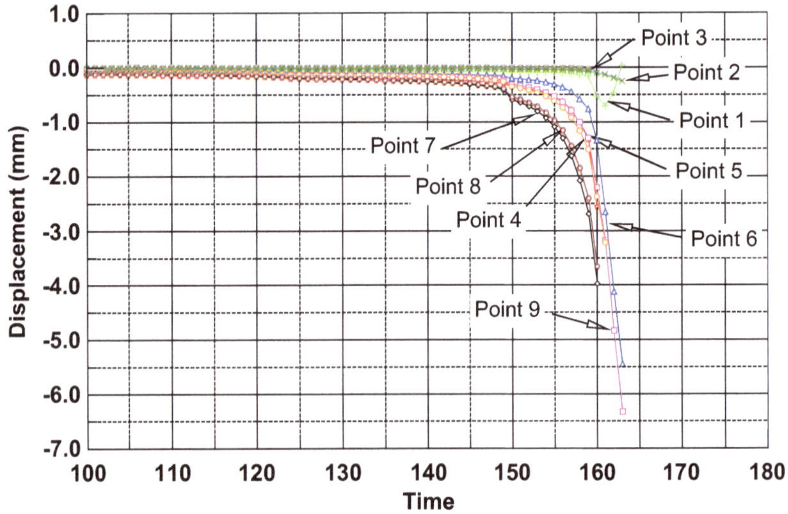


Fig. 17.4 Vertical displacement history at structural nodes

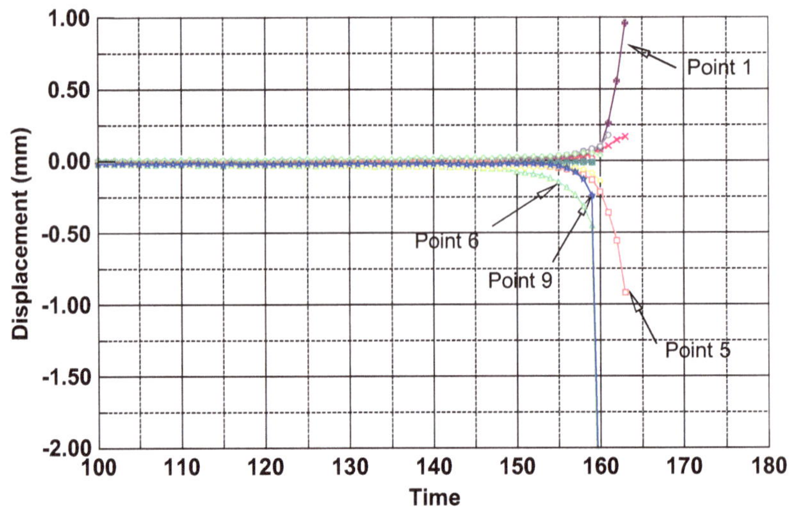


Shown in Figs. 17.3 and 17.4 were horizontal and vertical displacement histories at structural nodes 1–9. When load increasing, the displacements increased gradually. At the loading step 150, it was found that the horizontal displacement experienced large increment at nodes 4–9, especially at the nodes 7 and 8. Different with horizontal displacement, at this step relative large displacement could only been observed at the node 6 in vertical directions. After passing step 150, the structure moved faster. The horizontal displacement increased quickly at the nodes 4–9. It indicated that the step 150 is "instable threshold". Between the steps 159 and 160, large vertical displacement could be seen at the nodes 6, and large horizontal displacements were also found at nodes 4–9. After that, structure moved rapidly toward ground and left, finally fail.

In order to find out the location of potential failure with displacement measurements, a new concept—Bar Strain (1) Bar Normal Strain and (2) Bar Shear Strain were introduced. Bar normal strain ($\varepsilon$) refers to length change of a unit length of a truss between two connected nodes. Bar shear strain ($\gamma$) refers an angle rotation of the truss. For a regularly truss structural system, the structural strain could be calculated by:

$$\text{For horizontal truss: } \varepsilon = \frac{\Delta u}{L}, \gamma = \frac{\Delta v}{L} \quad \text{For vertical truss: } \varepsilon = \frac{\Delta v}{L}, \gamma = \frac{\Delta u}{L}$$

Where $\Delta u$ and $\Delta v$ are components of the relative displacement of two nearby structural nodes in x and y directions, L is the distance between the two nodes.

Shown in Figs. 17.5 and 17.6 were strain histories of the bars in the tested temporary structure. The strains were positive in bars 7–8 and 5–8, and negative in other bars. In other words, the bars 7–8 and 5–8 were in tension, others were in

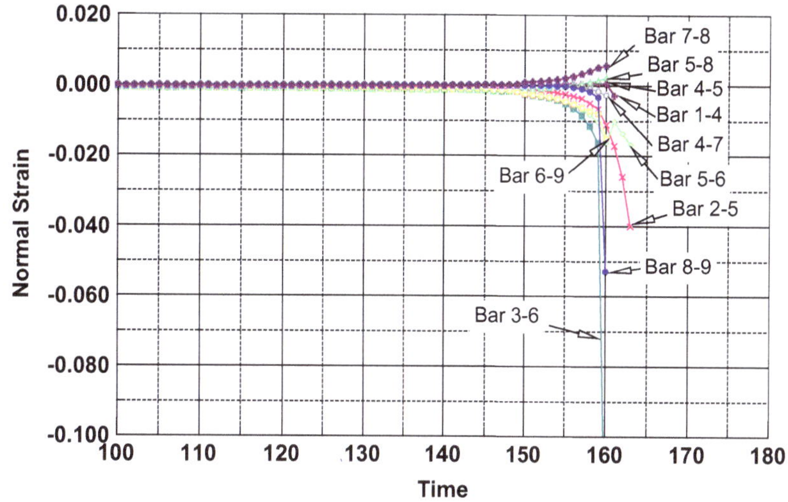


**Fig. 17.5** Normal strain histories of structural bars

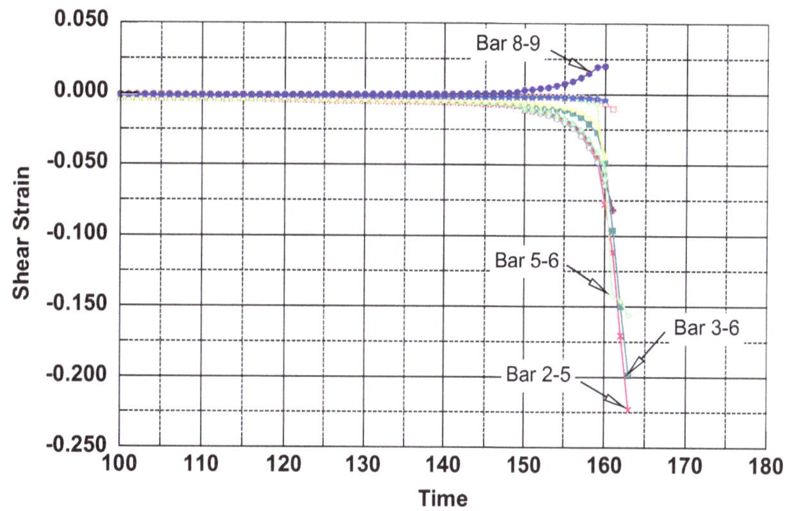


**Fig. 17.6** Shear strain histories of structural bars

compression. For most of the bar (except the bar 3–6 and 6–9), there were no significant normal strain before the time step 150. The magnitude of normal strain in the bar 3–6 and 6–9 did increase but slowly during the loading process. Beginning from the time step 150, the magnitude of the bar normal strains increased faster in the bars 2–5, 3–6 and 6–9. After the time step 159, the magnitude of normal strain rapidly increased in the bar 3–6, which indicated the failure of the structure. At this step, the magnitude of normal strain had increased 1.6 %. In contact, the magnitude of normal strain increased in the bar 5–8 and 7–8 before failure. As shown in Fig. 17.6, there was notable shear strain (rotation of the bar) in the bar 2–5, 5–6 and 3–6 starting from the step 120. The magnitude of shear strain increased slowly until the step 150. After this step, the magnitude of shear strain increased faster in the bar 2–5, 5–6 and 3–6, as well as the bar 6–9. The large shear strains indicated rotations of local structure. Consider the large normal strain increment in these bars, it could be concluded that local failure will occur in the bar 2–5, 5–6 and 3–6 zone. The failure initially could be in bar 3–6, which was consistent with the result of the experiment.

## 17.4 Conclusion

Instability test was conducted for a small scale temporary structure. Digital Image Correlation was utilized for structural monitoring and evaluation of the temporary structure during loading period. A concept of *Bar Strain* was introduced for structural instability analysis. The experimental results demonstrated the displacement and strain measurement of DIC could be used to determine the safety stage of the structure, predict the structural failure time and location. This state of the art measurement technology potentially could facilitate a real-time automated monitoring and detection system for of structural failures of temporary structures, and contribute to efforts to develop efficient design codes, standards, policies, and regulations to guide the construction and management of temporary structures on building sites.

## References

1. Ayub, M.: Structural collapses during construction: lessons learned. Structure Magazine, 12–14 December (2010)
2. Kleinosky, T.J.: Temporary structure failure—case studies—collapses from 2008 to present. https://failures.wikispaces.com/Temporary+Structure+Failure+−+Case+Studies.
3. Feld, J., Carper, K.L.: Construction Failure, 2nd edn. Wiley, New York (1996)
4. Yuan, X.: Overview of temporary structures failures in construction case studies from 1955 till present. http://failures.wikispaces.com/Overview+of+Temporary+Structures+Failures+in+Construction
5. Times square renovation thrills some, leaves others longing for old-time grittiness. http://www.cnn.com/TRAVEL/NEWS/9801/01/times.square/.
6. Sutton, M.A., Wolters, W.J., Peters, W.H., Ranson, W.F., McNeill, S.R.: Determination of displacements using an improved digital correlation method. Image Vis. Comput. **4**(3), 143–150 (1983)
7. Sutton, M.A., McNeill, S.R., Jang, J., Babai, M.: The effects of sub-pixel image restoration on digital correlation error estimates. Opt. Eng. **27**(3), 173–175 (1988)
8. Bruck, H.A., McNeill, S.R., Sutton, M.A., Peters, W.H.: Digital image correlation using Newton Raphson method of partial differential correction. Exp. Mech. **29**, 261–267 (1989)
9. Sutton, M.A., Bruck, H.A., Chae, T.L., Turner, J.L.: Development of computer vision methodology for the analysis of surface deformations in magnified images. ASTM STP-1094 on MICON-90, Advances in Video Technology For Microstructural Evaluation of Materials, vol. 109 (1990)
10. Ning, J., Xu, S., Wang, Y., Lessnerb, S.M., Sutton, M.A., Anderson, K., Bischoff, J.E.: Deformation measurements and material property estimation of mouse carotid artery using a microstructure-based constitutive mode. J. Biomech. Eng. **132**, 121010 (2010)
11. Xu, S.: Failure detection of temporary structures with digital image correlation for construction safety applications. In: 2015 International Digital Image Correlation Society Conference & Workshop, Columiba, SC USA, Nov 2015
12. Tiwari, V., Sutton, M.A., McNeill, S.R., Xu, S., Deng, X., Fourney, W.L., Bretall, D.: Application of 3d image correlation for full-field transient plate deformation measurements during blast loading. Int. J. Impact Eng. **36**(6), 862–874 (2009)
13. Xu, S., Tiwari, V., Deng, X., Sutton, M.A., Fourney, W.L.: Identification of interaction pressure between structure and explosive with inverse approach. Exp. Mech. **51**(6), 815 (2011)

# Chapter 18
# Using Digital Image Correlation to Detect Cracking in Opalinus Shale

**Stephen Philip Morgan**

**Abstract** The extraction of hydrocarbons from unconventional oil and gas reservoirs relies on a detailed understanding of the cracking processes in shale. Also, underground structures designed for nuclear waste repositories are often preferred to be in shale due to its characteristic low permeability. Specifically, how cracks initiate, propagate and coalesce in shale is of interest in these contexts. A series of unconfined compression tests were conducted on Opalinus shale extracted from the Mont Terri underground rock laboratory in Switzerland. These tests were performed on prismatic Opalinus shale specimens with one vertical flaw. High resolution imagery was used to capture crack initiation, -propagation and -coalescence. Digital Image Correlation (DIC) methods were used to determine deformation and crack initiation. By using static loading and unloading cycles, localized strains associated with cracking events were observed. The DIC analysis showed much more cracking than visual inspection with the naked eye. Additionally, a significant influence of bedding planes on the cracking patterns was observed.

**Keywords** Shale • Rock fracture • Crack detection • DIC • Cyclic loading

## 18.1 Introduction

The extraction of hydrocarbons from unconventional oil and gas reservoirs relies on a detailed understanding of the cracking processes in shale. Also, underground structures designed for nuclear waste repositories are often preferred to be in shale due to its characteristic low permeability. Specifically, how cracks initiate, propagate and coalesce in shale is of interest in these contexts. However, one of the biggest problems with experimental testing of rock is accurately measuring localized strains leading to crack initiation and propagation. For example, many numerical models require localized strength parameters that are analogous to the micro-scale mechanical behavior, such as the critical opening- and shear strains at the tip of a crack. Previously, experimental techniques included crack mouth opening devices (CMOD) and surface mounted strain gauges. However, these methods require pre-existing knowledge of the exact location of the crack initiation and of the propagation direction. Image processing techniques (Digital Image Correlation) have made it possible to measure surface displacements and calculate localized strain fields from imagery captured throughout the test.

There have been many experiments using cameras to determine the initiation, propagation direction, and mode of cracking in rock associated with flaw pairs [1–3] ("flaws" refer to rectangular openings pre-cut into intact rock). Also, there have been similar experiments conducted on the cracking in sedimentary rock such as Opalinus shale [4]. More recently, DIC techniques have been applied to tests conducted on rock [5–7]. The main purpose of many of these studies was to determine the existence of a fracture process zone before visual crack propagation or to detect cracking at the sub-pixel scale. In this study the DIC method will be applied to images taken during a uniaxial tests on Opalinus shale specimens with pre-cut flaws.

S.P. Morgan (✉)
Massachusetts Institute of Technology, 3 Portsmouth Street Apt. 2, Cambridge, MA 02141, USA
e-mail: spmorgan@mit.edu



A.T. Zehnder et al. (eds.), *Fracture, Fatigue, Failure and Damage Evolution, Volume 8*,
Conference Proceedings of the Society for Experimental Mechanics Series, DOI 10.1007/978-3-319-42195-7_18

## 18.2 Experimental Setup

The purpose of the experimental setup in this study was to observe cracking in Opalinus shale specimens under uniaxial compression and then associate these cracking events with stress-strain-time data. The specimens were uniaxially loaded in a ~270 kN capacity Baldwin™ hydraulic loading frame. Steel brush end platens were used on the specimen boundaries to reduce end effects. A photograph of the test setup is shown in Fig. 18.1. The tests were load-controlled at a constant load rate of ~5.3 kN/min. Load, axial displacement and time data were recorded from transducers in the Baldwin™ loading frame (refer to Fig. 18.1). High resolutions images were taken periodically (every 2 s) during the test to capture deformation and cracking. These still images were taken with a Canon™ 6D high resolution camera with a 100 mm Tokina™ lens.

The material used in this study, Opalinus shale, was provided by the Mont Terri Underground Laboratory in Switzerland. Opalinus shale is a clay based shale in the Dogger formation occurring in the Jura Mountains and other areas in Switzerland, France and Southern Germany. One of the advantages of using Opalinus shale is that it has been extensively tested (both in-situ and in the laboratory) and is a very well documented material [8–11].

The prismatic specimens tested were 10 cm × 5 cm × 2.5 cm (see Fig. 18.2). The edges of the specimens were cut with a standard tabletop band-saw with specialized carbide-tipped saw blades. A flaw oriented in the vertical direction, with a length of 12 mm and a width of 1 mm, was cut by drilling small holes at the flaw tips and then inserting a thin removable scroll saw blade to connect the flaw tips. Drill bits of 0.660 mm (#71) and 0.787 mm (#68) were used. Rectangular scroll saw blades with 0.381 mm thickness of and a 0.813 mm width were used to cut the flaws. Opalinus shale is a sedimentary rock with natural weak planes created during formation (referred to as bedding planes). The bedding planes in all of the specimens were oriented in the horizontal direction (perpendicular to the loading direction). Since Opalinus shale has a high clay content, it can be very sensitive to desaturation. Therefore, extra care was taken to preserve and store these specimens to reduce any saturation loss. Specimens were vacuum sealed and placed in secondary storage containers to preserve in-situ saturation.

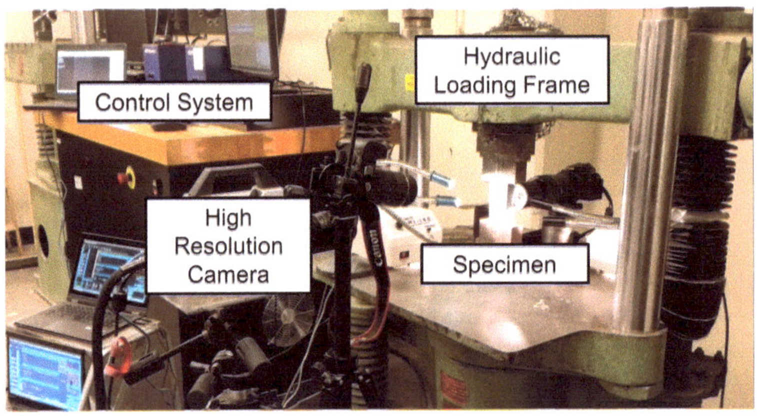


**Fig. 18.1** Photograph of the test setup in the current study. The specimens were hydraulically loaded under uniaxial load. A high resolution camera periodically took photos throughout the test (one image every 2 s)

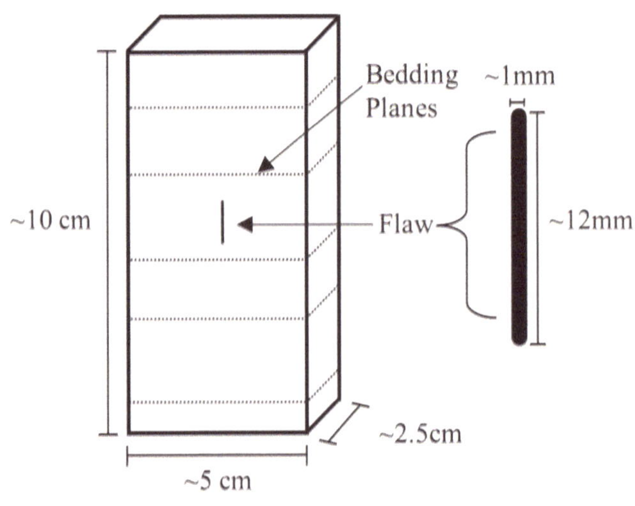


**Fig. 18.2** Schematic of the specimen configuration. The Opalinus shale specimen tested in this study was prismatic (10 cm × 5 cm × 2.5 cm) with a 12 mm long pre-cut flaw. The naturally formed bedding planes were oriented perpendicular to the loading direction (spacing variable)

**Table 18.1** Parameters of the captured images and of the NCORR DIC analysis

| | |
|---|---|
| *Image parameters* | |
| Resolution | 3648 × 5472 (20 megapixels) |
| Frame frequency | Every 2 s |
| Bit depth | 24 |
| *NCORR DIC parameters* | |
| Subset window radius | 20 pixels |
| Subset spacing | 2 pixels |
| Strain radius | Five displacement points |

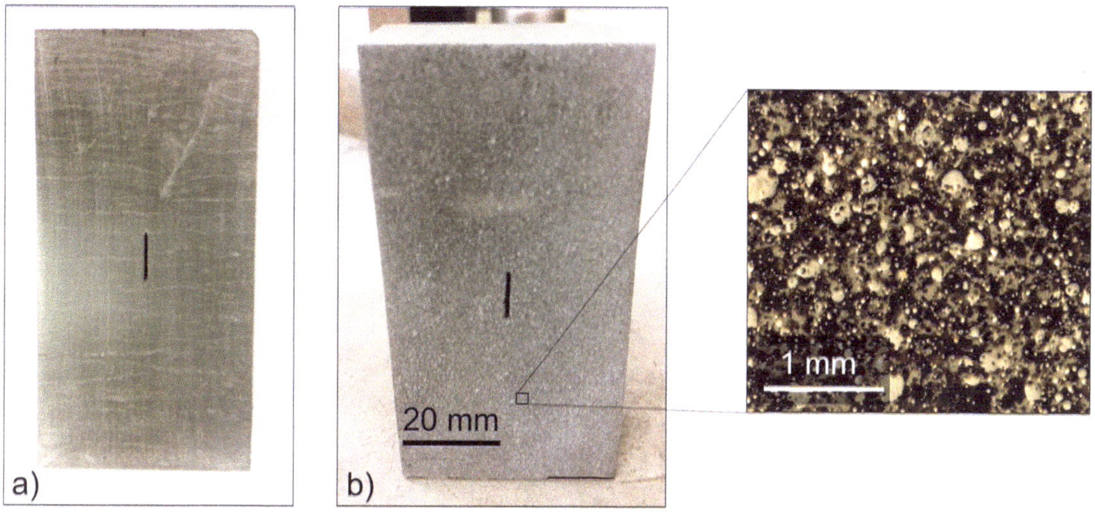


**Fig. 18.3** Speckle pattern on shale specimens. A shale specimen (**a**) without a speckle pattern and (**b**) with a black and white speckle pattern applied to the surface. The speckle pattern was applied using black and white spray paint

Three specimens were tested in this study. One test [A] was loaded statically until approximately 18 kN (13 MPa). The second test [B] was loaded to approximately 4.5 kN (3.5 MPa), unloaded to 0 kN and then loaded to 18 kN (13 MPa). The third test [C] was loaded to 4.5 kN (3.5 MPa) and unloaded to 0 kN three times, and then finally was loaded to 18 kN (13 MPa).

The DIC analysis was conducted using the NCORR code developed for Matlab™ [12]. The parameters used in NCORR are presented in Table 18.1. To improve the pattern detection with the DIC technique, a speckle pattern was painted on the surface of the specimens (Fig. 18.3). Black and white spray-paint was used (refer to Fig. 18.3). Speckle sizes range from 10 to 400 μm in diameter.

## 18.3 Results

Here we will discuss the results of the DIC analysis and the progression of strain localization observed in Opalinus shale under uniaxial loading. With the particular loading and flaw configuration cracks will occur as tensile cracks (opening) in the vertical direction (parallel to the loading direction). Therefore, only the strains in the x-direction ($E_{xx}$) are analyzed in this study. For more information on how strains were calculated the reader should refer to the NCORR algorithm documentation [12].

The results of the DIC analysis can be seen in Fig. 18.4. For all tests, minor strain localizations were observed at low applied stresses (3.0 MPa) while extensive strain localizations were observed at higher applied stresses (13.0 MPa) (Refer to Fig. 18.4). Any small difference in strain localization patterns between the tests (A–C) was likely due to variability of the material and specimen preparation. Since Opalinus shale is a natural material, different specimens can have different densities and locations of micro-cracks and bedding planes (weak zones). It also appeared that the amount of cycles had little to no effect on the strain patterns.

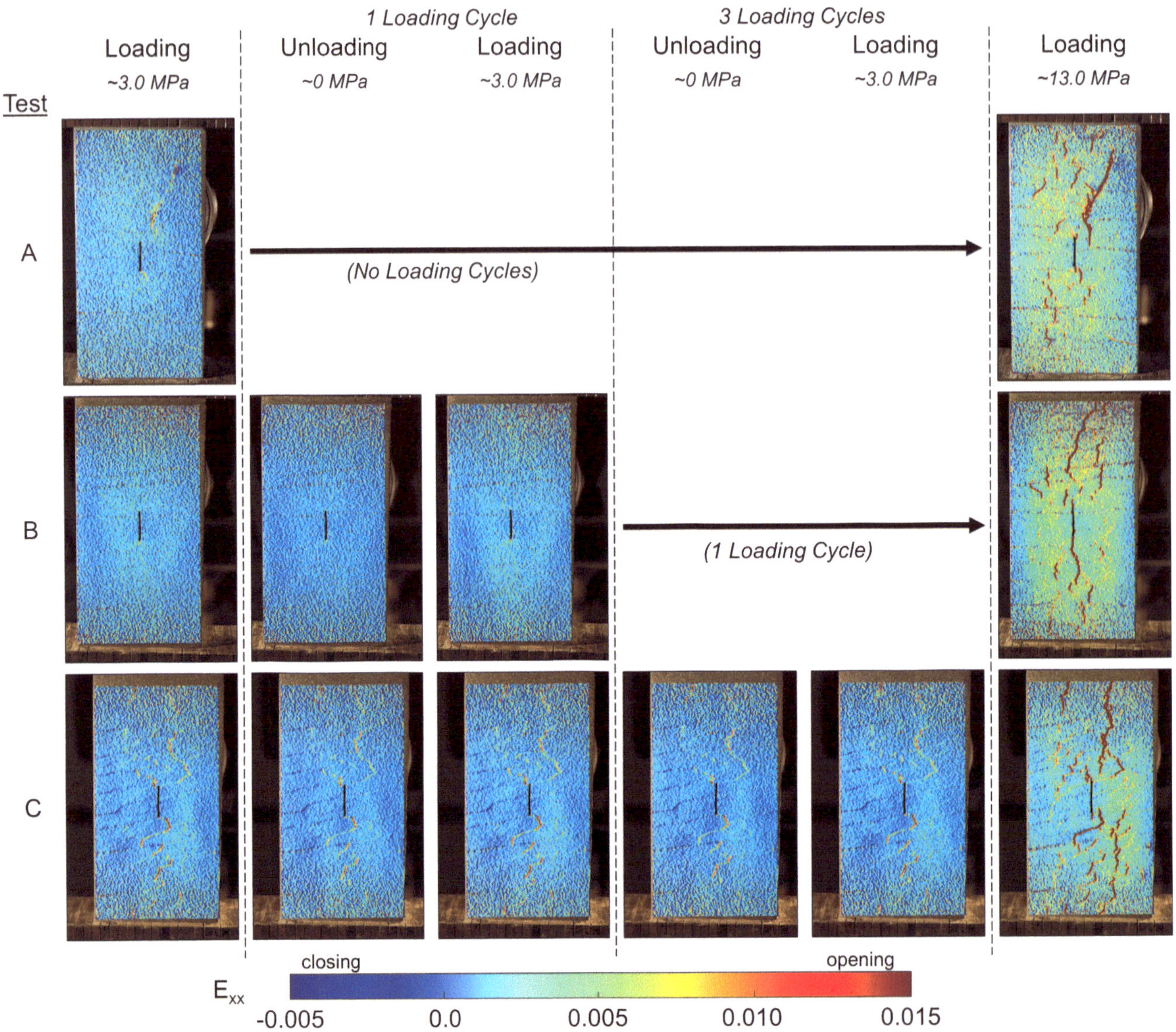


**Fig. 18.4** Progression of the lateral strain ($E_{xx}$) determined by the DIC analysis for all three tests. Test A was conducted with no loading cycles. Test B was conducted with one loading cycle from 0 to 3.0 MPa. Test C was conducted with three loading cycles from 0 to 3.0 MPa. All three tests were loaded to a final (pre-failure) stress of 13.0 MPa. Loading cycles appeared to have little to no effect on the observed strain ($E_{xx}$). All tests showed significant strain localizations at higher stresses

**Table 18.2** The strain in the x-direction ($E_{xx}$) averaged over a 5 cm × 5 cm area in the middle of the specimen

| | Average $E_{xx}$ ($\times 10^{-3}$) | | | | | |
|---|---|---|---|---|---|---|
| | | Cycle 1 | | Cycle 3 | | |
| | Initial load | Unload | Load | Unload | Load | Final load |
| | ~3.0 MPa | ~0 MPa | ~3.0 MPa | ~0 MPa | ~3.0 MPa | ~13.0 MPa |
| A | 1.32 | – | – | – | – | 4.78 |
| B | 1.18 | 0.80 | 1.19 | – | – | 4.57 |
| C | 1.11 | 0.90 | 1.08 | 0.87 | 1.05 | 3.95 |

The strain was averaged over the middle of the specimens to avoid noisier strains caused by out of focus pixels at the specimen ends and the effect of end constraints (refer to Fig. 18.4)

For each image the strain was averaged over a 5 cm × 5 cm area at the center of the specimen to avoid noisier strains caused by out-of-focus pixels at the specimen ends and the effect of end constraints. The resulting average lateral strains are presented in Table 18.2. From these results, it can be seen that there was essentially no increase in average strain as the

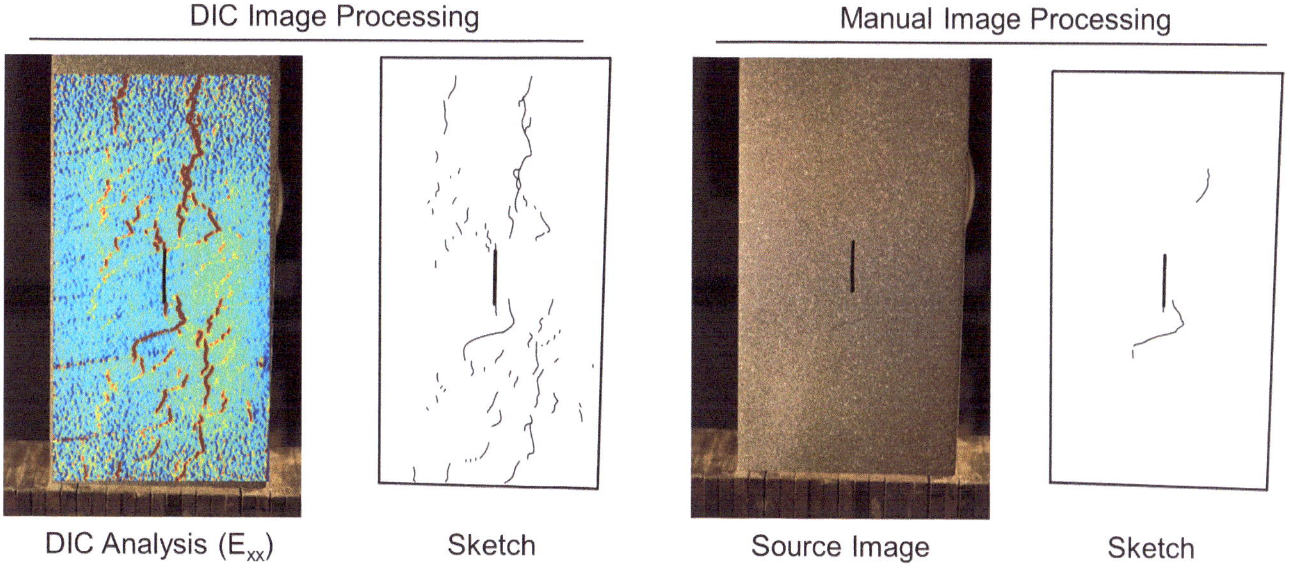


**Fig. 18.5** Comparing the DIC crack detection and the naked eye crack detection. Shown on the left is the DIC analysis ($E_{xx}$) of test C at ~13.0 MPa along with a sketch of the cracks associated with large strain concentrations (>0.010). The source image associated with this DIC analysis and a sketch of the cracking observed with the naked eye is shown on the right. The DIC analysis is capable of identifying much more cracking than naked eye inspection

number of cycles increases (refer to Table 18.2). In the third test (C), there was actually a slight decrease in the lateral strain. Although essentially no change in strain occurred up to three loading cycles, it is possible that a strain increase would have been observed with more loading cycles.

At higher stresses (~13 MPa) significant strain localization was observed in all tests (refer to Fig. 18.4). It should be noted that the overall strain concentration patterns observed were similar in all tests. This included some strain concentrations at the flaw tips and many smaller strain concentrations located throughout the specimen. Most interesting was the large amount of smaller strain concentrations away from the flaw tips. It is believed that strain concentrations produced by the flaw should cause the cracks to initiate at the flaw tips first and then propagate into the specimen. The results showed however, that although there were high strain concentrations at the flaw tips, there were many other strain concentrations that simultaneously occurred elsewhere in the specimen.

It is highly likely that these strain localizations observed in the results are related to cracking. However, little cracking was observed when using naked eye visual analysis. Since the opening of a discontinuity causes a strain gradient to occur, a crack will appear as a large localization of strain, and the width of this strain localization will be effected by the area of strain interpolation (i.e. strain window or smoothing). For this study a threshold strain value of 0.010 ($E_{xx} > 0.010$) was used to identify sub-visual cracks. By using this strain threshold, much more cracking was observed with the DIC analysis than with naked eye observations (Fig. 18.5). It is important to note that any cracks that were observed in the visual images (naked eye) corresponded to locations of strain concentrations in the DIC analysis (refer to Fig. 18.5). One can therefore conclude that the other locations of high strain are also cracks, and that these cracks could not be observed with naked eye inspection (It should be noted that this testing technique is only capable of observing the surface of the specimen and that there is currently no way of determining the depth of cracks into the specimen.)

Some of the cracks detected using the DIC analysis were clearly affected by the presence of naturally occurring bedding planes (Fig. 18.6). In some cases cracks followed bedding planes, however, in many cases cracks appeared to propagate from or arrest at a bedding plane (refer to Fig. 18.6). This behavior created en-echelon cracking arrays which could be observed most clearly in test C where the bedding planes were not perfectly horizontal. This slight inclination of the bedding planes produces some shear strains under uniaxial loading. Shear strains along the bedding planes could be the cause of these s-shaped en-echelon cracks observed in the DIC results for test C (refer to Fig. 18.6).

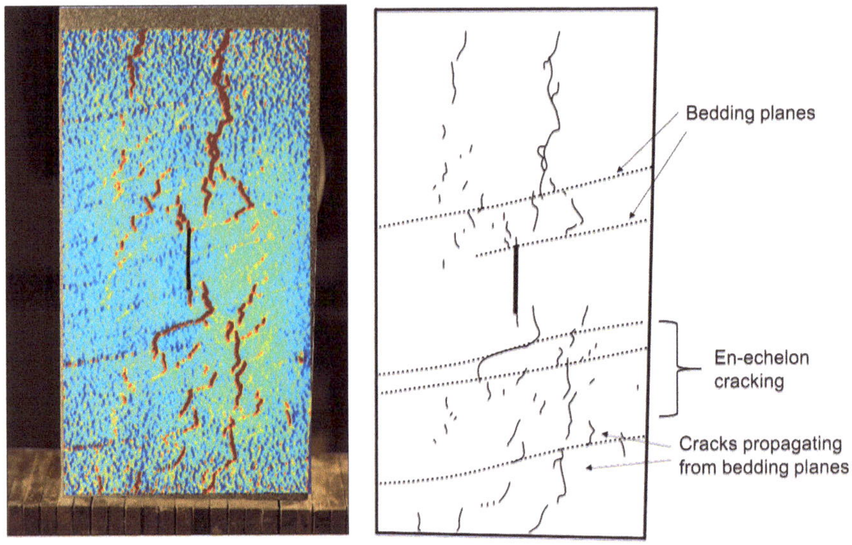


**Fig. 18.6** Crack-bedding plane interaction. The DIC results of test C at ~13.0 MPa are shown on the left and a sketch of the associated cracking, along with the approximate locations of visible bedding planes, is shown on the right. Interruptions in opening (tensile) cracks were observed near the bedding planes. Also, en-echelon cracking was observed

## 18.4 Summary and Conclusions

This study presented the DIC analysis results of three uniaxial compression tests conducted on Opalinus shale with pre-cut flaws. The results can be summarized as follows:

- More cracks were observed in Opalinus shale using the DIC method compared to visual image inspection with the naked eye
- Cracks observed with the DIC method appeared to be influenced by the presence of bedding planes
- Cycling loading had no observable effect on the lateral strain calculated from the DIC analysis

The ability of DIC to observe small cracks in rock material, which would otherwise not be seen using visual image inspection, is very important for rock testing. Many rock testing techniques such as compression-, Brazilian-, three-point notch bending-, and direct tension tests typically use imaging techniques to determine crack initiation and propagation. However, rock being a natural material with variable strength causes cracking patterns to be very complex, especially in layered anisotropic rocks such as shale. Therefore, the results of this study, which show that many cracks determined using the DIC method cannot be observed visually with the naked eye, are very encouraging for future shale testing.

Also very interesting is the clear effect of bedding planes on the cracking patterns observed with DIC. This includes cracks along and near bedding planes and en-echelon arrays between bedding planes. The influence of bedding planes on the cracking mechanisms is still not fully understood and can be very hard to capture. This level of detail of the crack-bedding plane interaction has not been observed in Opalinus shale tests before.

**Acknowledgements** The research presented in this paper was supported by TOTAL SA. Their support and advice is gratefully acknowledged. We would also like to thank Swisstopo and the Mt. Terri Laboratory (Dr. Ch. Nussbaum) for making the Opalinus shale cores available to us.

## References

1. Wong, L.N.Y., Einstein, H.H.: Crack coalescence in molded gypsum and carrara marble: part 1—macroscopic observations and interpretation. Rock Mech. Rock Eng. **42**(3), 475–511 (2009)
2. Wong, L.N.Y., Einstein, H.H.: Systematic evaluation of cracking behavior in specimens containing single flaws under uniaxial compression. Int. J. Rock Mech. Min. Sci. **46**(2), 239–249 (2009)
3. Morgan, S.P., Johnson, C.A., Einstein, H.H.: Cracking processes in barre granite: fracture process zones and crack coalescence. Int. J. Fract. **180**, 177–204 (2013)
4. Morgan, S.P., Einstein, H.H.: The effect of bedding plane orientation on crack propagation and coalescence in shale. In: 48th U.S. Rock Mechanics Symposium, Minneapolis, Minnesota, 1–4 June 2014
5. Lin, Q., Labuz, J.F.: Digital image correlation and the fracture process in rock. In: 44th U.S. Rock Mechanics Symposium and 5th U.S.–Canada Rock Mechanics Symposium, Salt Lake City, Utah, 27–30 June 2010

6. Nguyen, T.L., Hall, S.A., Vacher, P., Viggiani, G.: Fracture mechanisms in soft rock: identification and quantification of evolving displacement discontinuities by extended digital image correlation. Tectonophysics **503**, 117–128 (2011)
7. Zhao, C., Bao, C., Zhao, C., Tian, J: Experimental study on the damage and failure properties of rock-like material with pre-existing double flaws under uniaxial compression. In: Proceedings of the 13th International Congress of Rock Mechanics, Montreal, Canada (2015)
8. Bock, H.: Mont Terri project. RA experiment: rock mechanics analyses and synthesis; data report on rock mechanics. Mont Terri Consortium, Technical Report TR 2000-02 (2001)
9. Bock, H.: Mont Terri project. RA experiment: updated review of the rock mechanics properties of the Opalinus clay of the Mont Terri URL based on laboratory and field testing. Q+S Consult, Technical Report TR 2008-04 (2009)
10. Amann, F., Button, E.A., Evans, K.F., Gischig, V.S., Blumel, M.: Experimental study of the brittle behavior of clay shale in rapid unconfined compression. Rock Mech. Rock. Eng. **44**, 415–430 (2011)
11. Bossart, P., Burrus, F., Nussbaum, C.: Techniques de forages et d'excavations dans les argiles á Opalinus, laboratoire souterrian du Mont Terri. Swiss Bull. Angew. Geol. **17**(1), 3–28 (2012)
12. Blaber, J., Adair, B., Antoniou, A.: Ncorr: open-source 2d digital image correlation matlab software. Exp. Mech. **55**(6), 1105–1122 (2015)

# Chapter 19
# Early Detection of Damage Mechanisms in Composites During Fatigue Tests

**D. Palumbo, R. De Finis, P.G. Demelio, and U. Galietti**

**Abstract** Conventional procedures and methods used for evaluating the fatigue performance of materials require time consuming tests with high number of specimens. In the last years different methods have been developed with the aim to reduce the testing time and then the cost of the experimental campaign. In this regard, thermographic methods represent a useful tool for rapid evaluating the fatigue damage and the fatigue limit.

This work puts forward a novel procedure for assessing the fatigue limit and monitoring damage in GFRP material by means of thermography. In view of this, the proposed method allows for inferring several information related to the damage phenomena in composite materials, and, moreover, the reported results show a good agreement with those provided by the conventional procedures.

**Keywords** GFRP composites • Thermography • TSA • Fatigue damage • Fatigue limit

## 19.1 Introduction

Composite materials are nowadays exploited in the field of the production of large components for boating-yachting, aeronautical or aerospace industry [1]. A practical application of composites involves in wind turbine blades that are made from polymer composites since it is required high specific stiffness, strength, and good mechanical behavior [2]. Referring to mechanical characterisation, to perform the fatigue tests according to the Standards test methods, requires experimental campaigns in laboratory on sample specimens or directly on large components. This classical procedures for evaluating fatigue limit of material results in expensive and time consuming tests because of the high number of specimens being tested [3].

In the last years, with the aim to reducing testing times and costs of fatigue tests, different techniques and methods have been proposed to study rapidly and consistently the various damage phenomena [4–7]. In particular, the Infrared Thermography Technique (IRT) has been used by many authors to investigate fatigue damage in metallic and composite materials [8–10] and several methods and data analysis have been developed to obtain the information about the fatigue behaviour by the study of the heat sources generated during tests [11–15].

Luong [16] and Risitano [17] proposed a graphical method to assess fatigue limit of metallic materials by monitoring the superficial temperature of the specimen during a stepwise loading procedure. The same approach has been employed in Montesano work [11] for determining the fatigue limit of polymer matrix composite (PMC). In this case, the entire stress-life curve has been determined by means IRT and it was obtained an excellent correlation with the conventional lifespan curve. Another author, Steinberger et al. [12], adopts two different approaches (passive and active) to perform a quantitative characterization of damage by calculation of the loss factor via the hysteretic heating.

By exploiting thermography, other authors [18, 19] use a specific data processing of recorded infrared sequences to investigate the damage phenomena in the material. In this case, the temperature signal is analysed in the time domain therefore the first and the second order harmonics of the signal can be used to describe the nonlinear signal contents in the temperature evolution, due to the thermomechanical coupling and dissipation phenomena. This approach has been used in the work of Colombo et al. [20] to obtain the fatigue curve of GFRP specimens made of E-glass/epoxy with a 50 % volume fibre content. Also in this case, IRT provides a good estimation of the material life in the finite life region.

In several works [21–24] the potential to identify small damage by means of Thermoelastic stress analysis (TSA) has been demonstrated.

D. Palumbo • R. De Finis (✉) • P.G. Demelio • U. Galietti
Department of Mechanics, Mathematics and Management, Politecnico di Bari, Bari, Italy
e-mail: davide.palumbo@poliba.it; rosa.definis@poliba.it; giuseppepompeo.demelio@poliba.it; umberto.galietti@poliba.it



A.T. Zehnder et al. (eds.), *Fracture, Fatigue, Failure and Damage Evolution, Volume 8*,
Conference Proceedings of the Society for Experimental Mechanics Series, DOI 10.1007/978-3-319-42195-7_19

Thermoelastic Stress Analysis (TSA) is a non-contact, full field technique that provides stress maps of a component subjected to dynamic loading [21–24]. This technique is based on the thermoelastic effect; in fact, a component subjected to dynamic loading produces a small and reversible temperature changes. In adiabatic and linear elastic conditions, these temperature changes are proportional to the first stress invariant. Procedures based on TSA have been developed in the last years for the damage monitoring of Standard specimens and welded joints made of metallic materials (steel, titanium an aluminium) [25, 26].

The potential of TSA on composite materials has been showed in the works of Emery et al. [13] and Fruehmann et al. [14]. The first investigates different polymer-matrix-composites with different laminate type while the second highlights the possibility to use the phase signal for fatigue damage, even at low stress.

In this work a novel procedure is shown for processing thermographic data capable of describing the fatigue behavior of GFRP composites and to rapidly evaluate the fatigue limit of material. The main strong point of the proposed method is the possibility to obtain by a single thermographic data analysis information about the heat dissipative sources and the thermoelastic heat source.

Different fatigue tests were carried out on five standard specimens made of GFRP composites material and each test has been monitored at regular interval with an cooled infrared camera. Finally, conventional fatigue test were also carried out in order to obtain a comparison with the proposed procedure.

## 19.2 Theory

Generally, during fatigue tests, two thermal effects are generated: thermoelastic heat sources and intrinsic dissipations. The first represents the well-known thermoelastic coupling while, intrinsic dissipation is thermodynamically irreversible and is due to the viscoelastic nature of the matrix material, matrix cracking, fibre fracture, and interface cracking/friction among others [11].

Under the hypotheses of adiabatic conditions, the temperature changes $\Delta T_{el}$ for orthotropic materials are related to the changes in the stresses in the principal material directions by the following expression:

$$\Delta T_{el} = -\frac{T_0}{\rho C_p}(\alpha_1 \Delta\sigma_1 + \alpha_2 \Delta\sigma_2) \tag{19.1}$$

where $\alpha_1$ and $\alpha_2$ are the coefficients of linear thermal expansion relative to the principal axes, $C_p$ is the specific heat at constant pressure, $\rho$ is the density, $T_0$ is the absolute temperature and $\Delta\sigma_1$ and $\Delta\sigma_2$ are the principal stresses.

Acquisition systems used in TSA provide usually a not radiometrically calibrated $S$ signal proportional to the peak-to-peak variation in temperature during the peak-to-peak variation of the sum of principal stress. $S$ is usually presented as a vector, where modulus is proportional to the change in temperature due to the thermoelastic effect and the phase $\varphi$ means the angular shift between the thermoelastic and the reference signal provided by loading machine [13]. In this case, the following equation can be used:

$$A^* S = (\alpha_1 \Delta\sigma_1 + \alpha_2 \Delta\sigma_2) \tag{19.2}$$

where $A^*$ is a calibration constant. Equation (19.2) can be expressed in time domain as follows:

$$s = \frac{S}{2}\sin(\omega t + \pi + \varphi) \tag{19.3}$$

where $s$ is the uncalibrated thermographic signal, $\omega$ is the angular velocity and $\varphi$ is the phase angle between temperature and loading signal. This angle depends on a number of parameters such as, for example, thickness of the painting or the grips of the loading machine. Whilst phase can slightly change through the area analysed for non-perfect homogeneity of the surface conditions, it remains locally constant in presence of linear elastic behaviour of material and thus, if adiabatic conditions are achieved. In presence of damage, non-linearity of thermoelastic signal and phase variations can be observed [14].

As shown in Eq. (19.3), the thermoelastic signal varies at the same frequency of the loading during the test.

It was demonstrated that the intrinsic dissipations occur at twice the frequency of mechanical loading and are two order lower than thermoelastic ones. Intrinsic dissipations are irreversible sources opposed to thermoelastic ones. The presence of irreversible phenomena affect temperature by determining an increase of mean temperature of the specimen. In particular, in presence of damage, the superficial temperature of the specimen increase, then it will tend to stabilize on a plateau value and, in the eventuality of failure occurring at a certain loading step, temperature will abruptly increase.

## 19.3 Experimental Set-Up

Twelve specimens were obtained from a laminate panel made of an epoxy-type resin reinforced with two internal layer of uniaxial glass fiber and two external layer of quadriaxial fiber of the type 0°/90°/+45°/−45°. Dimensions of specimens were obtained according to standard ASTM D 3039 that were 25 mm wide, length of 250 mm and thickness of 2.5 mm. All the specimens were tested on a MTS (model 370, 100 kN capacity) servo-hydraulic machine.

Seven specimens were tested with conventional procedure in order to obtain the S-N curve and estimate the fatigue limit. In Table 19.1 are shown the maximum stress applied adopting a stress ratio of 0.1 and a loading frequency of 7 Hz.

The same fatigue test parameter were used for the thermographic tests. In this case, as shown in Table 19.2, a load stepwise procedure was carried out starting by a nominal stress amplitude ($\Delta\sigma/2$) of 30 MPa. At the end of each step (about 10,000 cycles) of loading machine the applied load was increased according to values shown in Table 19.2.

The experimental set-up adopted for thermographic tests is shown in Fig. 19.1.

An IR cooled In-Sb detector FLIR X6540 SC (640 × 512 pixel matrix array, thermal sensitivity NETD < 30 mK) has been used either to collect the thermal data and for monitoring superficial temperature of specimens. In particular, thermal sequences were acquired during each loading step around to 8000 cycles in correspondence of steady-state temperature conditions. The adopted frame rate was 100 Hz. Each acquisition lasts 10 s, therefore 1000 frames were recorded. Thermal sequences were analyzed by software Matlab®.

**Table 19.1** Stresses and number of cycles to failure obtained on seven specimens

| Step | б max (MPa) | Number of cycles |
|---|---|---|
| 1 | 380 | 400 |
| 2 | 270 | 2630 |
| 3 | 240 | 13,521 |
| 4 | 200 | 52,434 |
| 5 | 175 | 120,540 |
| 6 | 150 | 351,588 |
| 7 | 138.5 | 1,189,803 |

**Table 19.2** Number of loading steps and correspondents applied stresses

| Step | Δσ/2 (MPa) | б min (MPa) | б max (MPa) | б mean (MPa) |
|---|---|---|---|---|
| 1 | 30 | 4 | 44 | 24 |
| 2 | 35 | 7 | 67 | 37 |
| 3 | 40 | 8 | 78 | 43 |
| 4 | 45 | 9 | 89 | 49 |
| 5 | 50 | 10 | 100 | 55 |
| 6 | 55 | 11 | 111 | 61 |
| 7 | 60 | 12 | 122 | 67 |
| 8 | 65 | 13 | 133 | 73 |
| 9 | 70 | 14 | 144 | 79 |
| 10 | 75 | 16 | 156 | 86 |
| 11 | 80 | 17 | 167 | 92 |
| 12 | 85 | 18 | 178 | 98 |
| 13 | 90 | 20 | 200 | 110 |
| 14 | 100 | 22 | 222 | 122 |

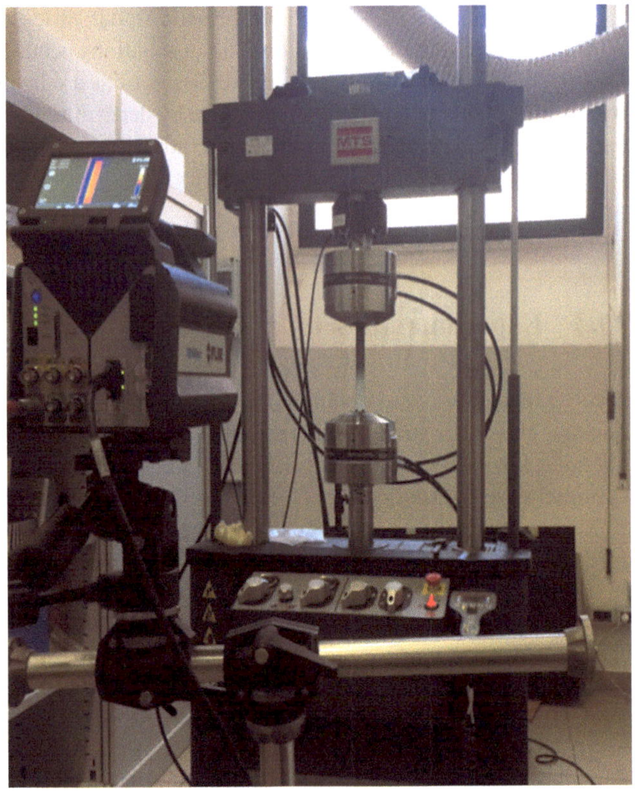


**Fig. 19.1** Experimental set-up adopted for thermographic tests

## 19.4 Methods and Data Analysis

A mathematical algorithm has been used to extract pixel by pixel information about the superficial temperature of the specimen in steady state conditions, the signal amplitude and the phase of the thermoelastic signal the amplitude of the second Fourier harmonic component related to the intrinsic dissipation. In particular, a suited thermographic signal model has been used to study the thermal signal $S_m$ evolution in the time domain, as indicate in Eq. (19.4):

$$S_m(t) = S_0 + at + S1 \sin(\omega t + \varphi) + S2 \sin(2\omega t) \tag{19.4}$$

where the term $S_0 + at$ represents the increase in mean temperature during the cyclic mechanical loading, $\omega$ is angular frequency of the mechanical imposed load, *S1* and $\varphi$ are respectively amplitude and phase of first harmonic component of Fourier series while, *S2* represents the amplitude of the second Fourier harmonic component. As just said, the term *S1* corresponds to the signal variation related to thermoelastic effect, *S2* term is proportional to the amplitude of intrinsic dissipation.

Equation (19.4) is integrated in the algorithm of software Matlab® providing image in form of data matrix for each constant parameter. The phase $\varphi$ of thermoelastic signal will be considered in further works.

The procedure for the processing of thermographic data was applied for each loading step and provides:

- the acquisition of the thermographic sequence. About 1000 frames were acquired for each sequence.
- assessing of $S_0$, $S_1$, $S_2$ pixel by pixel (Matlab® software),
- applying a Gaussian 2D-smoothing on data matrix obtained by Matlab® software,
- reducing data matrix to refer the analysis to only gauge length area. In this case, data matrix is the same for $S_0$, $S_1$ and $S_2$ (A1 area, Fig. 19.2).
- Subtracting the environmental temperature signal to the steady state temperature signal $S_0$ achieved during each step ($\Delta S_0$). Environmental temperature signal has been measured by using a dummy specimen (A2 area in Fig. 19.2). Evaluating $\Delta S_{0max}$ in the considered data matrix (A1 area).
- Subtracting a thermoelastic amplitude reference data matrix of the first loading step when there was no damage. In this way, the thermoelastic variations are compared to an undamaged reference condition ($\Delta S_1$).

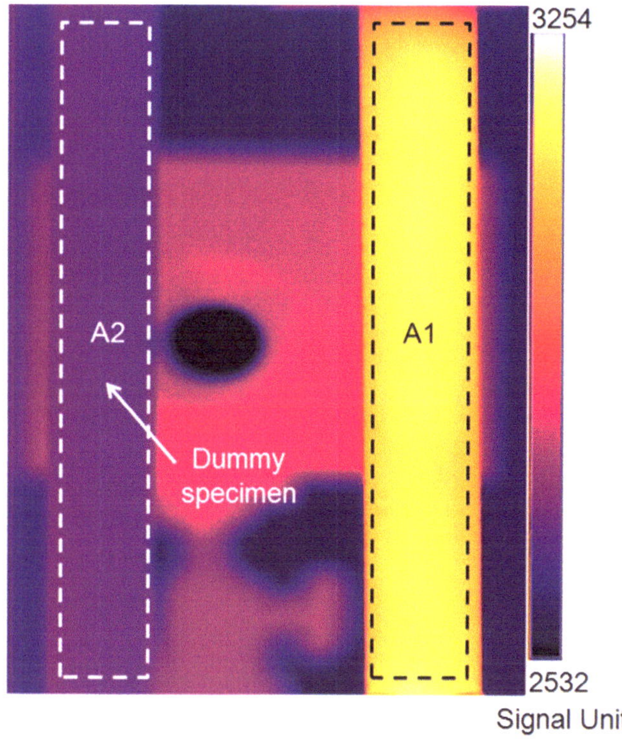


**Fig. 19.2** Area considered for the analysis (A1) and for evaluating of the environmental temperature signal (A2, dummy specimen)

Normalizing the thermoelastic data matrix with respect to the stress amplitude $\Delta\sigma/2$ in order to detect the thermoelastic signal variation only due to the damage ($S1_{norm} = \Delta S_1/\Delta\sigma/2$).

- Evaluation of the maximum and the minimum value of the thermoelastic signal ($\Delta S_1/\Delta\sigma/2$) in order to assess the $\Delta S1_{norm} = S1_{norm\_max} - S1_{norm\_min}$ from the phase data matrix. In order to avoid isolated bad pixels due for example, to "dead pixels", values of 98th and 2th percentile were used in this paper in place of $S1_{norm\_max}$ and $S1_{norm\_min}$ ($\Delta S1_{norm\_98\text{-}2perc}$).
- Evaluation of the maximum signal of the $S2$ signal ($S2_{max}$).

## 19.5 Results and Discussions

In Fig. 19.3 is shown the convectional curve *S-N* obtained with the data of Table 19.1 (seven specimens). This diagram allows for estimating of the fatigue limit of material for a conventional number of cycles. Considering a number of cycle of $2*10^6$ cycles, a value of 127.4 MPa is obtained for the fatigue limit in term of $\sigma_{max}$ and of 56.12 MPa in term of stress amplitude ($\Delta\sigma/2$).

In Fig. 19.4 the maps of thermoelastic signal are shown as the function of the applied stress amplitude for the specimen 1. It is clearly evident a signal variation in positive and negative value with respect to the reference stress condition of 30 MPa. As already demonstrated in other works [13, 14], TSA leads to localize the damaged areas of material. In particular, the thermoelastic signal variations are related to the redistribution of the stresses caused by the stiffness degradation due to the damage.

In Fig. 19.5 are shown the maps of the thermal signal at twice of the loading frequency (specimen 1). Also in this case, a significant increasing of the signal is obtained as the applied stress increases. These maps seem to provide different and complementary information about the type of the damage of material if compared with the thermoelastic maps. In this regard, further works are necessary to relate the different damage mechanisms to the thermographic analysis.

The trend of the maximum temperature signal measured in area A1 in steady state condition is shown in Fig. 19.6a. The signal increases for each loading step due to the viscoelastic nature of the matrix material until a significative increasing in correspondence of the presence of the damage mechanisms.

The procedure described in the work of De Finis et al. [15] has been used for evaluating the fatigue limit for each measured signal ($\Delta S_{0max}$, $\Delta S1_{norm\_98\text{-}2perc}$ and $S2_{max}$). For each specimen the adopted procedure consists in:

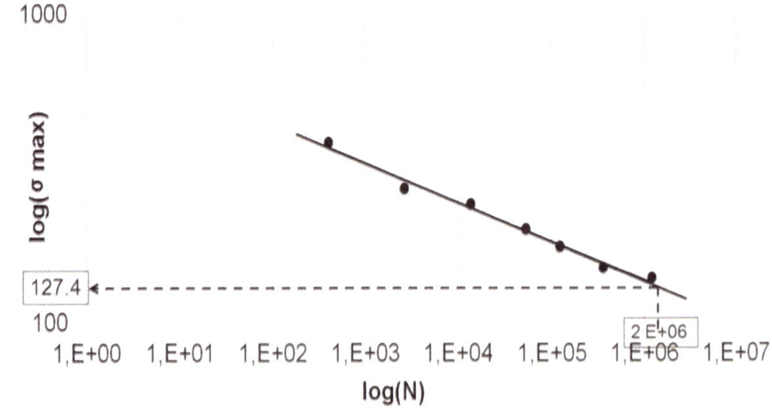


**Fig. 19.3** Conventional S-N curve and estimation of the fatigue limit in correspondence of a run-out limit of $2 \times 10^6$ cycles

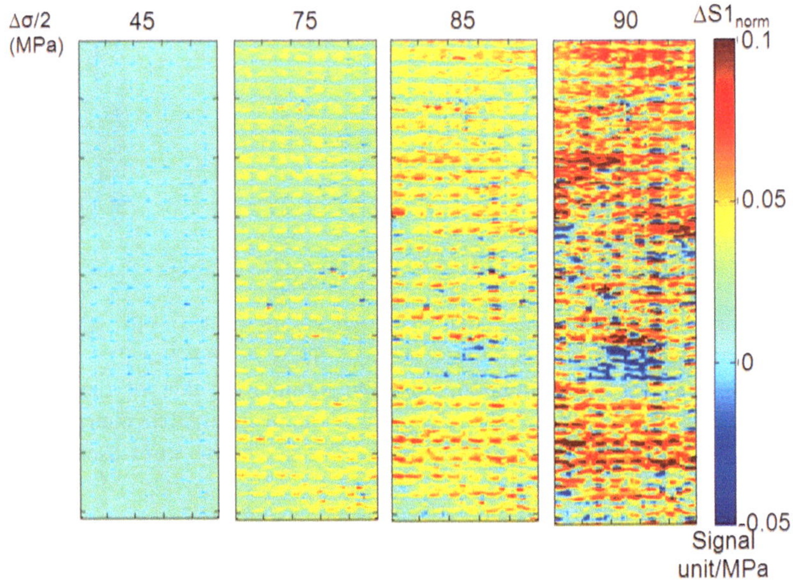


**Fig. 19.4** Maps of the thermoelastic signal obtained for four different loading step

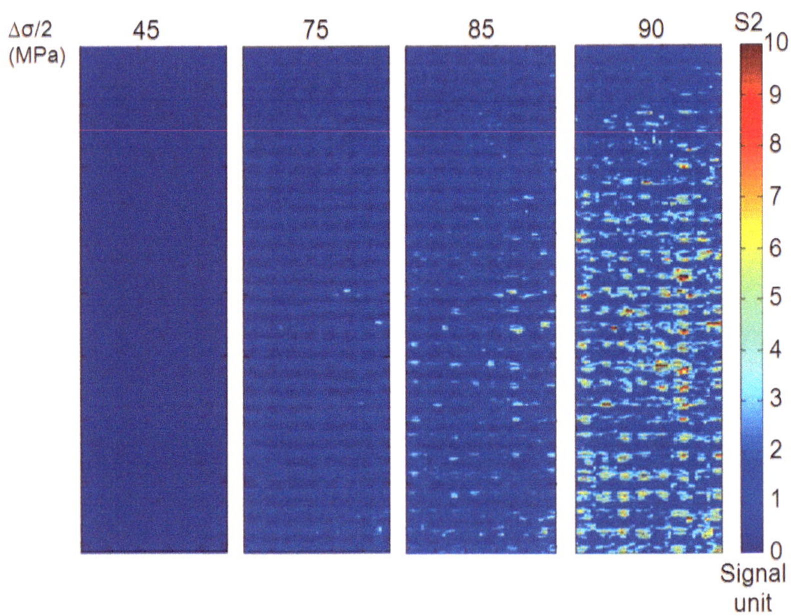


**Fig. 19.5** Maps of the thermographic signal at the twice of the loading frequency obtained for four different loading step

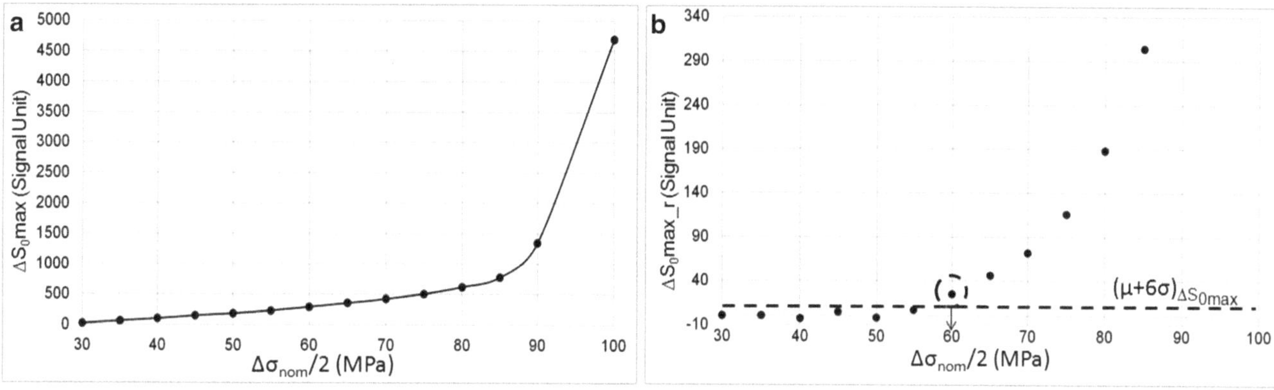


**Fig. 19.6** (**a**) Maximum thermographic signal vs. amplitude stress and (**b**) estimation of the fatigue limit with the method [15]

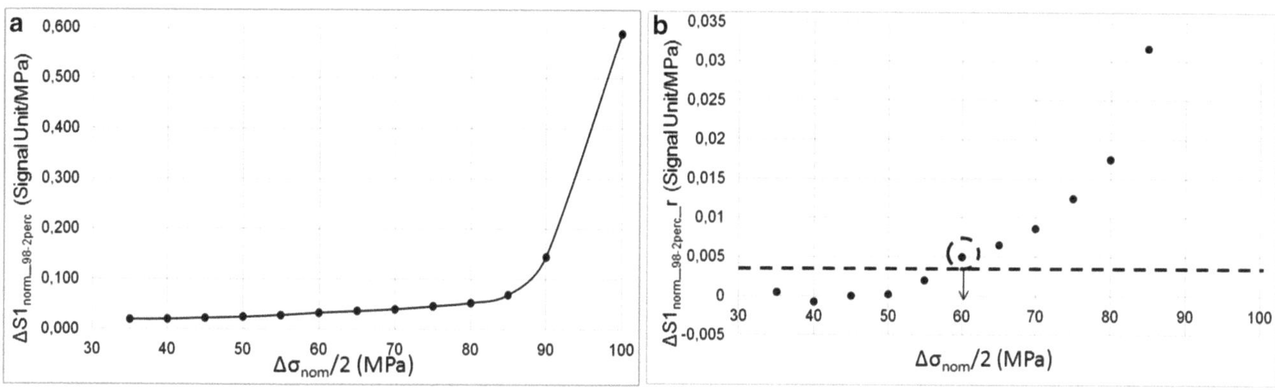


**Fig. 19.7** (**a**) Thermoelastic signal vs. amplitude stress and (**b**) estimation of the fatigue limit with the method [15]

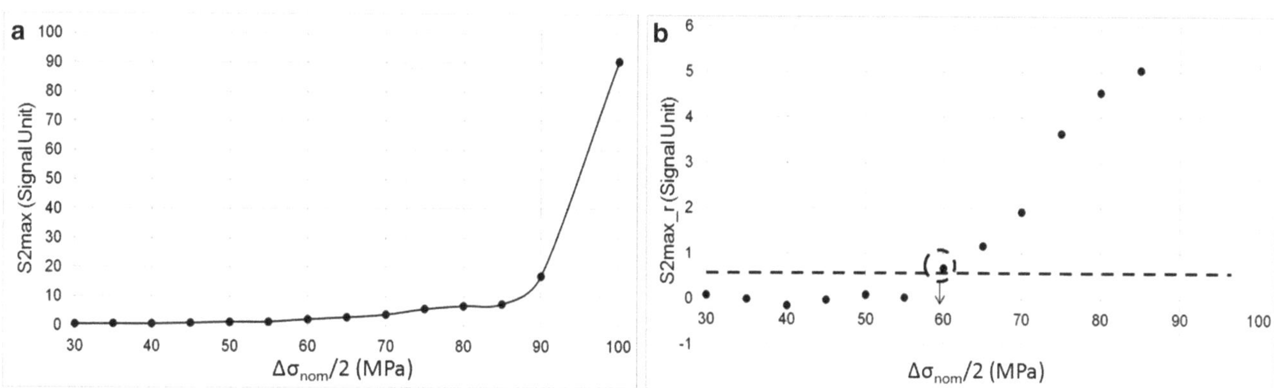


**Fig. 19.8** (**a**) Maximum thermographic signal at twice of loading frequency vs. amplitude stress and (**b**) estimation of the fatigue limit with the method [15]

1. Linear regression analysis of the first four data couples ($P$; $\Delta\sigma/2$) and evaluation of the best fit line ($y = mx + q$). $P$ represent the generic thermographic signal.
2. Evaluation of residuals of $P$ ($P_{\_r}$) for each loading step.
3. Evaluation of standard deviation ($\sigma_{P\_r}$) and mean ($\mu$) of residuals ($P_{\_r}$) of the first four data of each tests.
4. Evaluation of the threshold value $P_{th} = \mu + 6{*}\sigma_{P\_r}$
5. Evaluation of the first loading step (of $P_{\_r}$ data) for which the condition: $(P_{\_r})_N > P_{th}$ is verified (where $N$ is the number of the step). The first loading step exceeding the condition, is considered the estimation of fatigue limit.

Figures 19.6, 19.7 and 19.8b show in graphic way the procedure above exposed. In particular, the residual are plotted versus the stress amplitude and the dotted line represents the threshold value used for the estimation of the fatigue limit.

**Table 19.3** Summary of the results obtained with the thermographic technique and comparison with the conventional fatigue tests

| N° Specimen | Fatigue limit (MPa) | | | |
|---|---|---|---|---|
| | $\Delta S0_{max}$ | $\Delta S1_{norm\_98\text{-}2perc}$ | $S2_{max}$ | *S-N* curve ($2*10^6$ cycles) |
| 1 | 60.0 | 60.0 | 60.0 | **Δσ/2 = 56.12** |
| 2 | 60.0 | 65.0 | 60.0 | |
| 3 | 60.0 | 65.0 | 50.0 | |
| 4 | 65.0 | 70.0 | 55.0 | |
| 5 | 55.0 | 65.0 | 55.0 | |
| Average | **60.0** | **65.0** | **56.0** | |
| Standard deviation | **3.5** | **3.5** | **4.2** | |

In Table 19.3 are reported the fatigue limit values in terms of stress amplitude for each specimen and for each thermographic signal used for the analysis ($\Delta S0_{max}$, $\Delta S1_{norm\_98\text{-}2perc}$ and $S2_{max}$).

Results obtained with thermography technique are in good agreement with the fatigue limit obtained in conventional way by considering a run-out limit of $2*10^6$ cycles.

## 19.6 Conclusions

In this work a novel procedure has been proposed for evaluating the fatigue limit of GFRP composite materials with thermography technique. In particular, the uncalibrated signal has been analysed in the time domain in order to extract information about the thermal signal related to the thermoelastic and dissipative sources.

Five specimens were used for the fatigue tests and each specimen has been subjected to a loading step procedure until the failure. Thermal sequences were acquired during each step in correspondence of the steady state conditions. Three thermal signals were obtained from processed data: the thermal signal related to the increasing of the mean temperature of the specimen, the signal related to thermoelastic source and the signal related to the intrinsic dissipations.

A statistical method, validated for metallic materials has been used for evaluating the fatigue limit for each thermal signal extracted by the analysis. Results show a good agreement with those obtained by the conventional S-N curve.

Proposed procedure could represent a useful tool for the monitoring of real and more complex components subjected to actual loading conditions.

## References


1. Bannister, M.K.: Development and application of advanced textile composites. Proc. Inst. Mech. Eng. Part L: J. Mater. Des. Appl. **218**, 253–260 (2004)
2. Palumbo, D., Tamborrino, R., Galietti, U., Aversa, P., Tatì, A., Luprano, V.A.M.: Ultrasonic analysis and lock-in thermography for debonding evaluation of composite adhesive joints. NDT&E Int. **78**, 1–9 (2016)
3. Harris, B.: Fatigue in Composites. Woolhead, Cambridge (2003)
4. Munoz, V., Valès, B., Perrin, M., Pastor, M.L., Welemane, H., Cantarel, A.: Damage detection in CFRP by coupling acoustic emission and infrared thermography. Compos. Part B **85**, 68–75 (2016)
5. Goidescu, C., Welemane, H., Garnier, C., Fazzini, M., Brault, R., Péronnet, E., Mistou, S.: Damage investigation in CFRP composites using full-field measurement technique: combination of digital image stereo-correlation, infrared thermography and X-ray tomography. Compos. Part B **48**, 95–105 (2013)
6. Naderi, M., Kahirdeh, A., Khonsari, M.M.: Dissipated thermal energy and damage evolution of glass/epoxy using infrared thermography and acoustic emission. Compos. Part B **43**, 1613–1620 (2012)
7. Kordatos, E.Z., Aggelis, D.G., Matikas, T.E.: Monitoring mechanical damage in structural materials using complimentary NDE techniques based on thermography and acoustic emission. Compos. Part B **43**, 2676–2686 (2012)
8. Palumbo, D., Ancona, D., Galietti, U.: Quantitative damage evaluation of composite materials with microwave thermographic technique: feasibility and new data analysis. Meccanica **50**, 443–459 (2015)
9. Galietti, U., Dimitri, R., Palumbo, D., Rubino P.: Thermal analysis and mechanical characterization of GFRP joints. In: 15th European Conference on Composite Materials: Composites at Venice, ECCM 2012, Venice, Italy, 24–28 June 2012
10. Tamborrino, R., Palumbo, D., Galietti, U., Aversa, P., Chiozzi, S., Luprano, V.A.M.: Assessment of the effect of defects on mechanical properties of adhesive bonded joints by using non destructive methods. Compos. Part B **91**, 337–345 (2016)
11. Montesano, J., Fawaz, Z., Bougherara, H.: Use of infrared thermography to investigate the fatigue behavior of a carbon fiber reinforced polymer composite. Compos. Struct. **97**, 76–83 (2013)

12. Steinberger, R., Valadas Leitão, T.I., Ladstätter, E., Pinter, G., Billinger, W., Lang, R.W.: Infrared thermographic techniques for non-destructive damage characterization of carbon fibre reinforced polymers during tensile fatigue testing. Int. J. Fatigue **28**, 1340–1347 (2006)
13. Emery, T.R., Dulieu-Barton, J.K.: Thermoelastic stress analysis of the damage mechanisms in composite materials. Compos. Part A **41**, 1729–1742 (2010)
14. Fruehmann, R.K., Dulieu-Barton, J.M., Quinn, S.: Assessment of the fatigue damage evolution in woven composite materials using infra-red techniques. Compos. Sci. Technol. **70**, 937–946 (2010)
15. De Finis, R., Palumbo, D., Ancona, F., Galietti, U.: Fatigue limit evaluation of various martensitic stainless steels with new robust thermographic data analysis. Int. J. Fatigue **74**, 88–96 (2015)
16. Luong, M.P.: Infrared observation of thermomechanical couplings in solids. In: Thermosense XXIV Conference, Part of SPIE's Aerosense 1—Orlando, Florida, 5 Apr 2002
17. La Rosa, G., Risitano, A.: Thermographic methodology for the rapid determination of the fatigue limit of materials and mechanical components. Int. J. Fatigue **22**, 65–73 (2000)
18. Krapez, J.K., Pacou, D., Gardette, G.: Lock-in thermography and fatigue limit of metals. Quant. Infrared Thermography **6**, 277–282 (2000)
19. Ummenhofer, T., Medgenberg, J.: On the use of infrared thermography for the analysis of fatigue damage processes in welded joints. Int. J. Fatigue **31**, 130–137 (2009)
20. Colombo, C., Libonati, F., Pezzani, F., Salerno, A., Vergani, L.: Fatigue behaviour of a GFRP laminate by thermographic measurements. Procedia Eng. **10**, 3518–3527 (2011)
21. Harwood, N., Cummings, W.: Thermoelastic Stress Analysis. National Engineering Laboratory and Adam Hilger, New York (1991)
22. Pitarresi, G., Patterson, E.A.: A review of the general theory of thermoelastic stress analysis. J. Strain Anal. Eng. Des. **35**, 35–39 (1999)
23. Wang, W.J., Dulieu-Barton, J.M., Li, Q.: Assessment of non-adiabatic behaviour in thermoelastic stress analysis of small scale components. Exp. Mech. **50**, 449–461 (2010)
24. Palumbo, D., Galietti, U.: Data correction for thermoelastic stress analysis on titanium components. Exp. Mech. **56**, 451–462 (2016)
25. Palumbo, D., Galietti, U.: Characterization of steel welded joints by infrared thermographic methods. Quant. Infrared Thermography J. **11**(1), 29–42 (2014)
26. Galietti, U., Palumbo, D.: Application of thermal methods for characterization of steel welded joints. EPJ Web Conf. **6**, 38012 (2010)

# Chapter 20
# Implementing Noise, Multi-Frequency Stimulus, and Realtime Analysis to Nonlinear Model Tracking

**Timothy A. Doughty, Liam J. Cassidy, and Shannon M. Danforth**

**Abstract** Crack initiation in a cantilevered beam subject to harmonic excitation near the beam's second natural frequency has been determined using Nonlinear Model Tracking (NMT), a health monitoring technique. This method assumes a second order nonlinear differential equation model with cubic stiffness; the nonlinear parameter is tracked until catastrophic failure using a Continuous Time based System Identification. Previous research has shown that significant change in the value of the nonlinear parameter indicates the system's transition from healthy to unhealthy. This study introduced Gaussian noise into the raw stimulus and response data at various signal-to-noise ratios. The results were compared with those of the original data to highlight the technique's effectiveness in determining a change in the system's health. The model's robustness was also investigated by exciting the system at a range of frequencies near resonance, and the results of this test were compared to results from excitation at a single frequency. New methods of identifying crack formation in the beam were also implemented. The raw acceleration response data was plotted next to the nonlinear parameter in real time, and the system's natural frequency was recorded before and after crack initiation.

**Keywords** Nonlinear • Vibrations • Nondestructive health monitoring • Modal • Failure

## 20.1 Introduction

In all physical systems that undergo repetitive loading, damage due to fatigue is a major cause of system failure. Due to oscillatory loading, cracks tend to form at localized points of stress concentration [1, 7]. These cracks grow as the system continues to undergo repeated forcing and, if gone unnoticed, will cause a fracture and in turn, failure of the system.

Over the course of past decades researchers have developed Nondestructive Evaluation Techniques (NDE) for tracking failure of physical systems without causing catastrophic damage to the system [1]. Many conventional NDE techniques track shifts in a system's natural frequency using linear methods to detect crack initiation [2–6], and new methods for NDE are constantly being developed [8–10]. If the assumption is made that the system at hand can be modeled as linear, the system's health can be wrongly identified as failing due to the physical nature of the model's nonlinearity.

This study investigates the estimation of a nonlinear parameter in the equation of motion for a vibrating cantilevered beam as a means to detect crack formation. The cantilevered beam serves as a simplified model of a more complicated system such as an airplane wing, and a significant change in the nonlinear parameter can serve as a prediction of beam failure. The beam's slenderness pronounces the system's nonlinearities, and the response at the second mode has been chosen based on the shaker capabilities. Stimulus and response data are collected, and Continuous Time System Identification (CTSI) is used to find the nonlinear parameter [11]. As the system behavior transfers from healthy to unhealthy, the estimate of the cubic stiffness parameter changes dramatically. Mapping this nonlinear parameter over time serves as a robust way to monitor the beam's health.

T.A. Doughty (✉) • L.J. Cassidy • S.M. Danforth
University of Portland, 5000 N Willamette Blvd, Portland, OR 97203, USA
e-mail: doughty@up.edu; cassidy17@up.edu; danforth16@up.edu



A.T. Zehnder et al. (eds.), *Fracture, Fatigue, Failure and Damage Evolution, Volume 8*,
Conference Proceedings of the Society for Experimental Mechanics Series, DOI 10.1007/978-3-319-42195-7_20

## 20.2 Background

Previous research performed using NMT has shown that, while popular techniques based on linear modeling are much simpler, they can fail to accurately assess the health of a vibrating system because they ignores the healthy nonlinear behavior that may be present in a physical system [12]. NMT is able to account for both the simple linear behavior of the system as well as the healthy complex nonlinear behavior which all physical systems exhibit. As the system transitions from healthy to unhealthy, the nature of the system's nonlinear behavior changes, thus altering the nonlinear parameter in the model and indicating the onset of failure.

Also in previous studies, NMT has been able to accurately predict the onset of failure without identifying a physically accurate model or the parameters associated with that model. In past studies, the model has shown success with using a minimal amount of data, using a generic pre-collected data set, collecting data using strain gages instead of accelerometers, and testing various system geometries [13, 14]. Despite major changes in the system, the model's resiliency points to its robustness and real-world application.

For this study, the cubic stiffness parameter was chosen to be the nonlinear parameter tracked until fracture. The cubic stiffness nonlinearity simplifies the model by absorbing the other system nonlinearities and is robust with regard to small changes in the dynamics of the system [15].

The method used to collect data for tracking the change in the nonlinear parameter involves three stages. First, using National Instruments LabVIEW, the cantilevered beam is excited at a range of frequencies to determine an estimate of resonance. Next, healthy steady state data is measured at a series of frequencies near resonance. Lastly, the beam is excited at near-resonant frequency until failure. During the terminal break stage, the linear values previously solved for allow simple tracking of the nonlinear parameter.

## 20.3 Theoretical Model Development

In this study, a horizontally mounted slender cantilever beam is vertically excited near its second natural frequency. The theoretical configuration is shown in Fig. 20.1.

The model for the transverse displacement of a cantilevered beam with nonlinear bending stiffness developed in [16] is given with the following partial differential equation and associated boundary conditions:

$$m\ddot{v} + c\dot{v} + D_\xi\left\{v^{iv} + \left[v'\left(v'v''\right)'\right]'\right\} = -mA\Omega^2\cos(\Omega t), \tag{20.1}$$

$$v(0,t) = v'(0,t) = v''(L,t) = v'''(L,t) = 0. \tag{20.2}$$

When the excitation is modal, the displacement is known to be spatially dependent. Let $\phi(s)$ be the orthonormalized mode shape given by

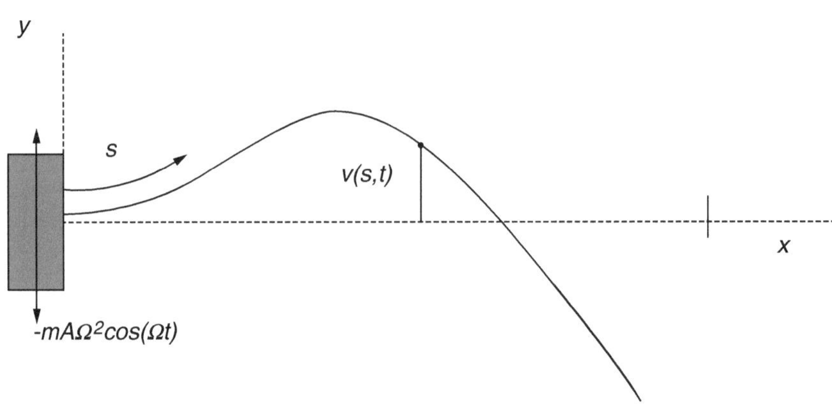


**Fig. 20.1** Theoretical configuration for base-excited cantilever beam, second mode shape

$$\phi(s) = C[\sin(\beta s) - \sinh(\beta s) - \Psi(\cos(\beta s) - \cosh(\beta s))] \tag{20.3}$$

Where:

$$\Psi = \frac{\sin(\beta L) + \sinh(\beta s)}{\cos(\beta L) + \cosh(\beta L)}. \tag{20.4}$$

For the second mode, as given in [17],

$$\beta L = 4.694. \tag{20.5}$$

Letting $v(s,t) = a(t) \cdot \phi(s)$, then the nonlinear differential equation of one variable appears as

$$m\ddot{a} + c\dot{a} + ka + \alpha a^3 = F(t) \tag{20.6}$$

Where:

$$K = \left\{ D_\xi \int_0^L \phi\phi^{iv} ds \right\}, \tag{20.7}$$

$$\alpha = \left\{ D_\xi \int_0^L \phi\phi''^3 + 4\phi\phi'\phi''\phi''' + \phi\phi'^2\phi^{iv} ds \right\} \tag{20.8}$$

And:

$$F(t) = -mA\Omega^2 \cos(\Omega t) \cdot \int_0^L \phi ds \tag{20.9}$$

This model resembles the classic mass-spring-damper model, but includes a single nonlinear cubic term. The model is reduced to the linear mass-spring-damper system in systems with weak nonlinearity.

## 20.4 Experimental Setup

The experimental setup is shown in Fig. 20.2. As in previous studies, a 6061 aluminum bar with dimensions 12.7 × 1.57 × 500 mm was fixed to an MB Dynamics 50 Modal Exciter. To excite the beam at a constant frequency, a sinusoidal function generator virtual instrument coded in LabVIEW was connected to an amplifier, which in turn relayed the signal to the shaker.

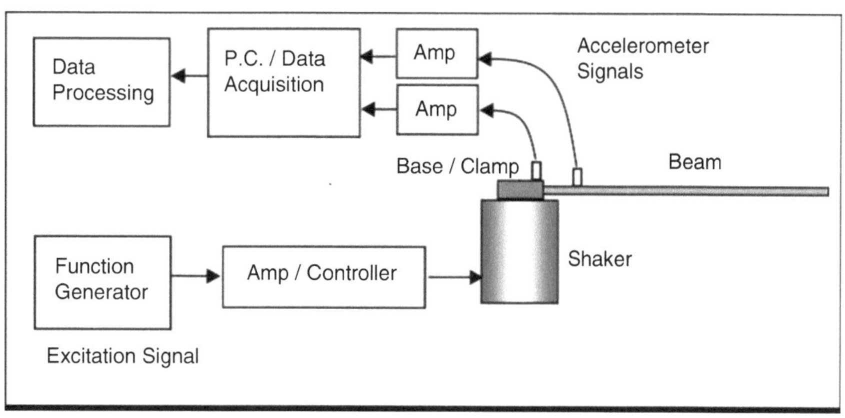


**Fig. 20.2** Experimental setup

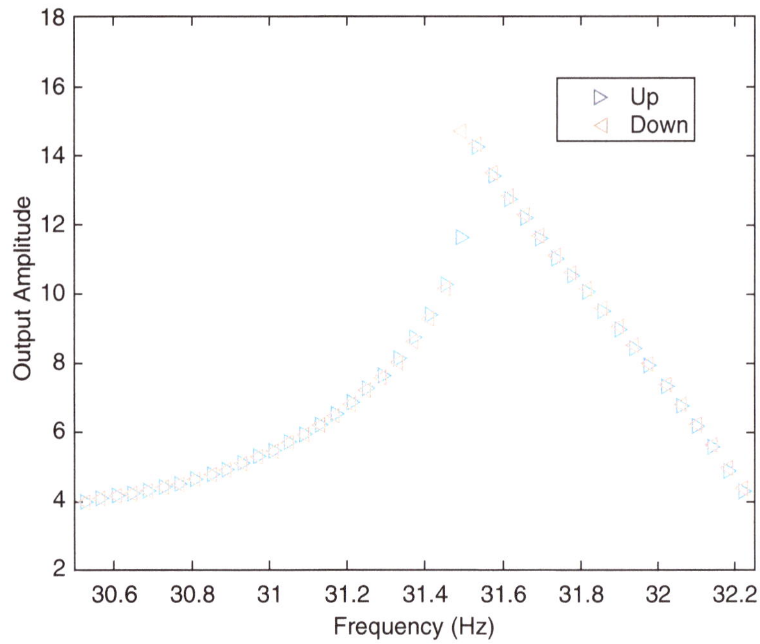


**Fig. 20.3** Example of nonlinear frequency response curve

Accelerometers were attached at the shaker platform and at 4.5 cm down the beam. Raw acceleration data from the accelerometers was collected by National Instruments data acquisition units.

First, acceleration data was collected during excitation at a range of frequencies. The beam's response was measured and plotted. The maximum response corresponded to the beam's second natural frequency. Figure 20.3 shows a typical frequency response curve of the system.

Once the beam's natural frequency was found, steady state data was collected at $\pm 1.5$ and $\pm$ 3 Hz from resonance in order to accurately identify the nonlinear model [18]. The steady state data was later used to create a frequency response fit and determine the linear parameters.

The beam was then excited at its second natural frequency until failure occurred. The accelerometers on the beam and shaker collected acceleration data throughout this stage. The data was imported into Matlab for post-processing. Realtime data processing was introduced in this study in order to increase the real-world application of the model and to allow for the testing of the beam's resonant frequency before and after crack formation.

The robustness of the model was investigated by adding noise into data during post-processing. Noise was added to the same data set at various signal-to-noise ratios (SNRs) to determine the noise threshold at which the model could still accurately indicate crack failure in the beam. White Gaussian noise was added to the raw stimulus and response data at SNR-values of 1.5, 10, and 15. See Fig. 20.4 for a time response of data with and without added noise.

This study also investigated the model's ability to handle multi-frequency stimulus. During processing, the raw response data is fit with the first 12 Fourier terms; these terms are integrated analytically to determine the velocity and position values. However, the Fourier fit is usually done with a single-frequency excitation. A swept sinusoidal signal was generated in LabVIEW and relayed to the shaker in order to excite the beam at a repeated range of frequencies near resonance to determine if a change in alpha could still be detected.

With the implementation of realtime analysis, the crack initiation in the beam was recognized immediately. This study investigated the change in the system's resonant frequency before and after crack formation. As always, at the beginning of the test, the system's response was recorded at a range of frequencies in order to determine the resonant frequency. This study included a second test for natural frequency performed again after a significant change in alpha.

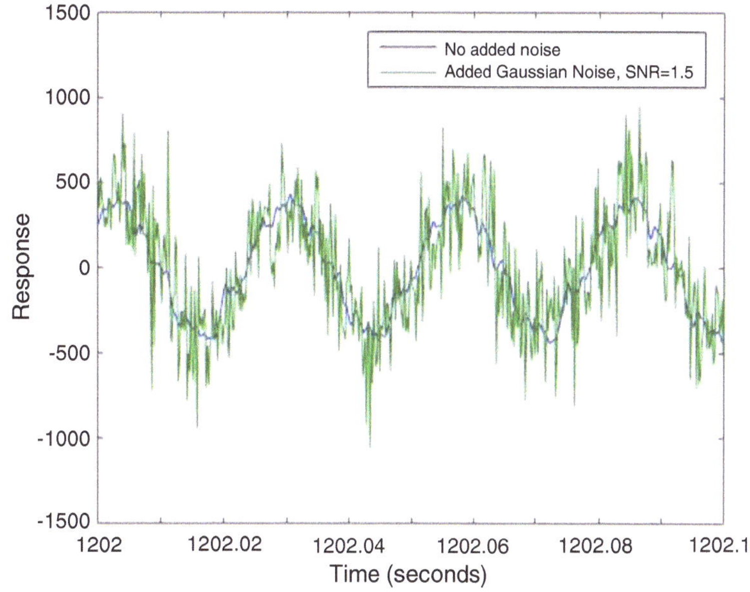


**Fig. 20.4** Time responses with and without added noise, SNR 1.5

## 20.5 System Identification

Healthy response data was collected at a range of frequencies surrounding the system's resonant frequency. The nonlinearity of the system was not pronounced during steady-state data collection due to a low amplitude of excitation, so the linear parameters could be back-solved for in a least-squares sense. The model has proven to be successful even though the resulting linear parameters have no physical meaning.

Once the linear parameters have been determined, the nonlinear parameter alpha can be estimated using the Continuous Time System Identification (CTSI) method. The model can be rewritten in matrix form as

$$\left[\ddot{a}\;\dot{a}\;a\;a^3\right]\left[m\;c\;k\;\alpha\right]^T = F(t) \tag{20.10}$$

with $\ddot{a}$, $\dot{a}$, $a$, and $F(t)$ representing position, velocity, acceleration, and forcing vectors, respectively. Solving this equation for $\alpha$ gives

$$\alpha = \{F(t) - m\ddot{a}(t) - c\dot{a}(t) - ka(t)\}\cdot\{a^3\}^T \tag{20.11}$$

The nonlinear parameter is then estimated with a least squares approximation.

## 20.6 Results

A plot of the nonlinear parameter alpha versus time, shown in Fig. 20.5, for a certain test specimen with no added noise resulted in a change in alpha after minute 200. A plot of the nonlinear parameter for the same test data with added noise at an SNR of 15, 10, and 1.5, respectively, can be seen in Figs. 20.6, 20.7, and 20.8.

As in past research, the nonlinear parameter was successfully tracked over the lifetime of a single-frequency excitation of the model. Figure 20.9 shows a plot of alpha for single-frequency excitation using realtime tracking. The plot of alpha does not resemble that of Fig. 20.5 because the data was processed using the method of system identification developed in [14]. In this plot, alpha gradually increases in value until minute 120, where it decreases dramatically. Processing the data and plotting alpha in realtime allowed the system's resonant frequency to be determined before and after crack formation. The resulting frequency response functions from this test are shown in Fig. 20.10. The beam's natural frequency was 34.9 Hz before the crack formed and 34.7 Hz after crack formation.

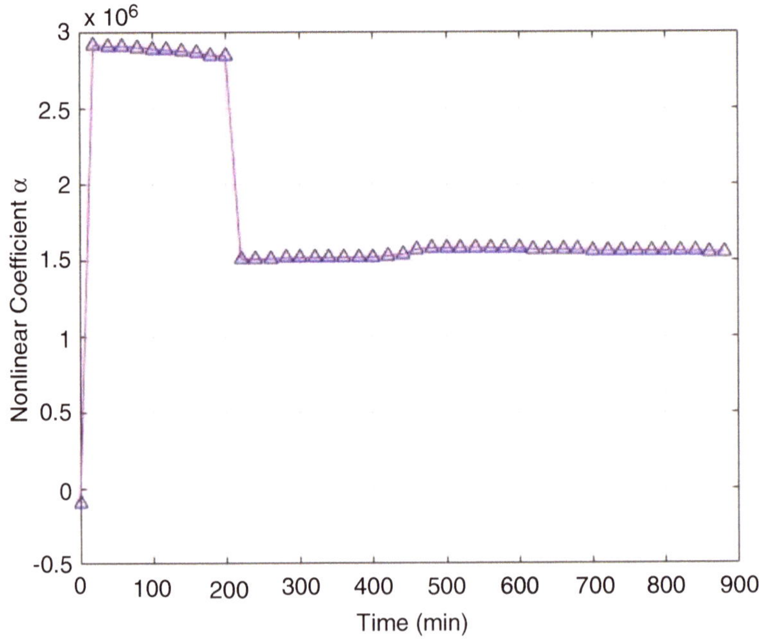


**Fig. 20.5** Alpha plot with no added noise

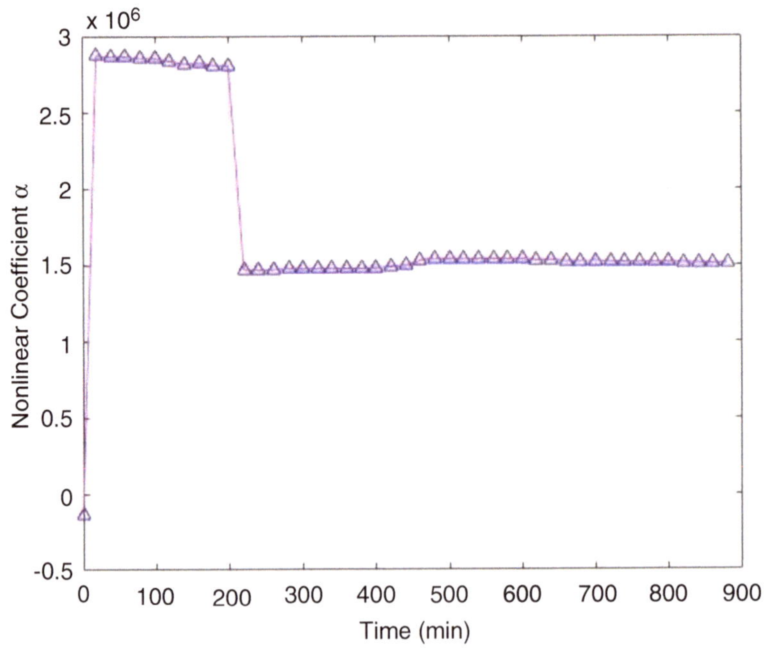


**Fig. 20.6** Alpha plot, SNR of 15

A final experiment investigated multi-frequency stimulus using the processing method from [14]. The raw acceleration data taken during the multi-frequency terminal shake stage was successfully imported and analyzed during live data processing. The plot of alpha versus time during a multi-frequency terminal shake can be seen in Fig. 20.11. Again, alpha appears to increase until crack formation around minute 75, where it begins to decrease.

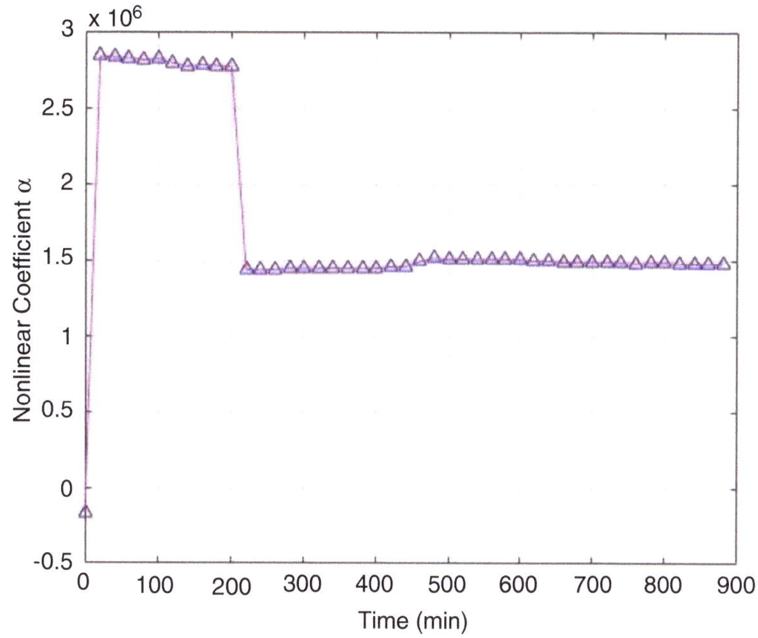


**Fig. 20.7** Alpha plot, SNR of 10

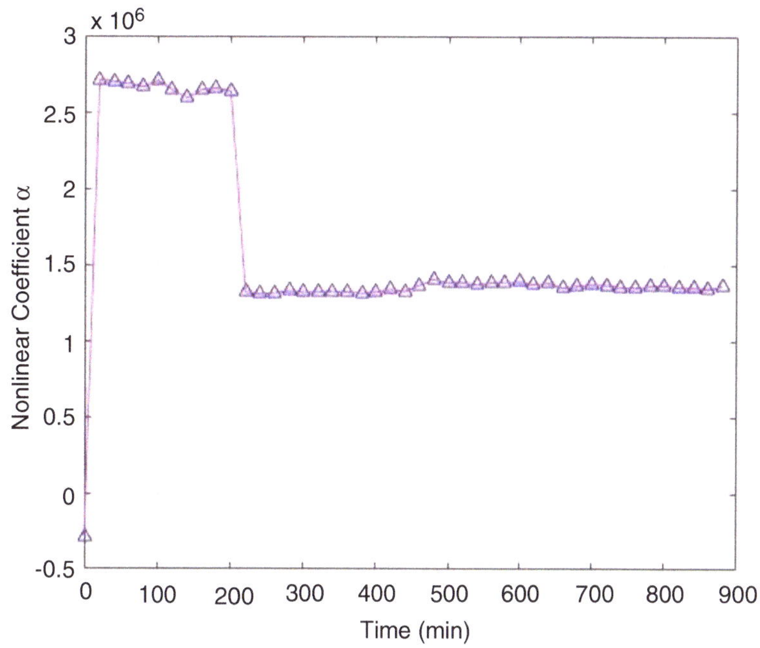


**Fig. 20.8** Alpha plot, SNR of 1.5

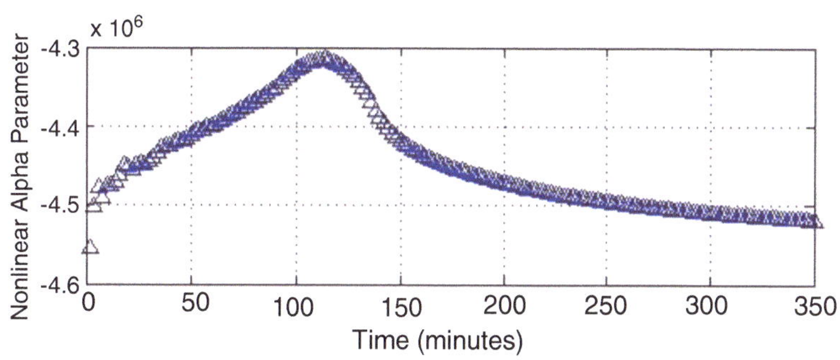


**Fig. 20.9** Alpha plot for single frequency excitation

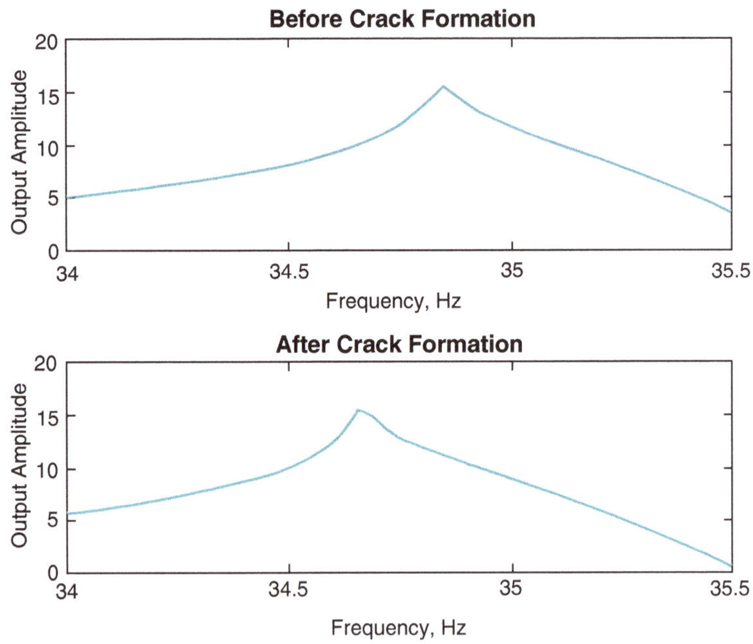


**Fig. 20.10** Beam's frequency response before and after crack formation

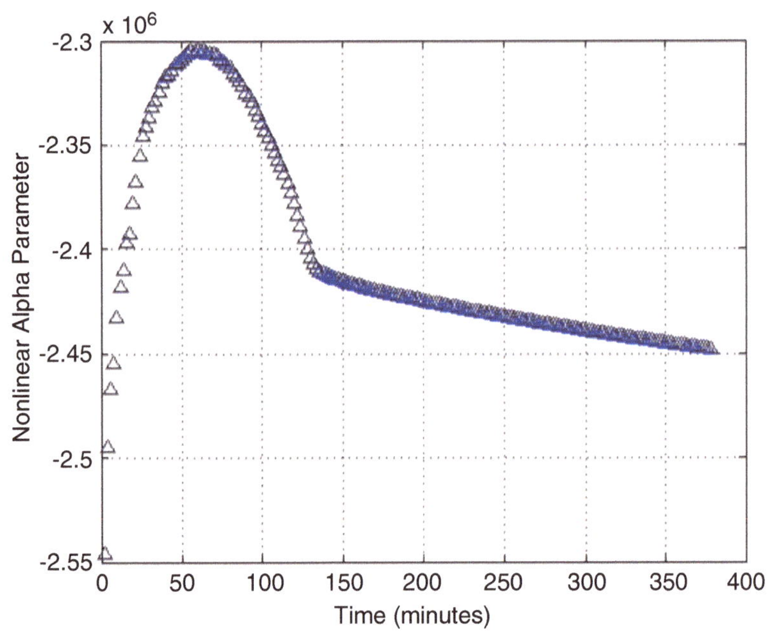


**Fig. 20.11** Alpha plot for multi-frequency excitation

## 20.7 Discussion

For the SNRs of 15, 10, and 1.5, the change in alpha occurred at roughly the same time as the data with no noise, at minute 200. The success in predicting crack formation despite the addition of Gaussian noise even at relatively large amplitudes is likely attributed to the method of post-processing the data, which includes Fourier fitting the stimulus and response signals in order to find the dominant frequency.

In previous studies, the system was excited at a single frequency during the failure stage of testing. Significant shifts in the value of alpha represented the shift from healthy to unhealthy mechanical behavior of the beam [10]. The use of the NMT method to track the mechanical health of the beam during multi-frequency excitation was proposed in order to enhance the robustness of the model. Most real world mechanical systems are not excited at strictly one frequency; the results of multi-frequency testing showed a significant shift in the nonlinear parameter after crack formation, thus suggesting that this study's NMT method applies to systems excited at a repeated sweep of frequencies near resonance as well.

As seen in Figs. 20.9 and 20.11, the nonlinear parameter shows a significant initial increase in value. This phenomena, termed early onset drift, has been encountered in previous research [13]. The cause of early onset drift of the nonlinear parameter is being investigated. However, the phenomena does not invalidate the method, as the behavior of the alpha parameter does change significantly after the period of early onset drift for both the single frequency and multi-frequency stimulation. An alternative explanation for the cause of this early rise in the alpha parameter is that the system was never able to exhibit healthy nonlinear vibration because it was in an unhealthy state to begin the terminal shake stage. This initial unhealthy behavior could be caused by physical damage to the test specimen, such as cracks on the surface or within the material.

The implementation of realtime tracking to this study allowed the beam's resonant frequency to be experimentally found before and after crack formation. A shift in the beam's natural frequency did occur as the system transitioned from healthy to unhealthy, but the difference was relatively small, only 0.2 Hz.

Linear health monitoring techniques track shifts in the system's linear parameters to indicate a shift in natural frequency, ignoring the manifestation of healthy nonlinear behavior [5]. Testing the beam's natural frequency before and after crack formation confirmed this hypothesis and showed that the NMT method was consistent with linear methods.

## 20.8 Conclusion

The NMT method was used to track mechanical health and predict the onset of failure for a cantilevered beam under oscillatory loading. The method detected the transition from healthy to unhealthy behavior of the system by tracking the nonlinear parameter alpha until catastrophic failure.

This study further promotes the use of Nonlinear Model Tracking for nondestructive health monitoring by proving the model's ability to track the mechanical health of systems with high noise levels present in measured signals. The model proved to be effective with signal-to-noise ratios as low as 1.5. The real-world application of the model is also enhanced due to the model's success with indicating crack formation during excitations swept over a range of frequencies. Furthermore, the realtime tracking allowed for direct observation of the system's failing health at times associated with changes in the estimated nonlinear parameter.

Though still only in the early stage of use, this NMT method shows promise for various health monitoring applications by demonstrating success with noisy data and multi-frequency stimulus and response. With the realtime tracking, more substantial mapping of the failing beam can be better correlated to behavior of the estimated nonlinear term. Further research should investigate the implementation of a larger range of excitation frequencies and amplitudes during multi-frequency excitation, including excitation at both the second and third modes, and transient excitation frequencies to increase the model's real-world application.

## References


1. Doebling, S.W., Farrar, C.R., Prime, M.B.: Summary review of vibration-based damage identification methods. Shock Vib. Dig. **30**, 91–105 (1998)
2. Khiem, N.T., Lien, T.V.: Multi-crack detection for beam by the natural frequencies. J. Sound Vib. **273**, 175–185 (2004)
3. Loutridis, S., Douka, E., Hadjileontiadis, L.J.: Forced vibration behavior and crack detection of cracked beams using instantaneous frequency. NDT&E Int. **8**, 411–419 (2005)
4. Lee, Y.Y., Liew, K.M.: Detection of damage locations in a beam using the wavelet analysis. Int. J. Struct. Stab. Dyn. **1**, 455–465 (2001)
5. Gudmundson, P.: Changes in modal parameters resulting from small cracks. Proc. Int. Model Anal. Conf. Exhib. **2**, 690–697 (1984)
6. Kim, J.-T., Ryu, Y.-S., Hyun-Man, C., Stubbs, N.: Damage identification in beam-type structures: frequency-based method vs. mode-shape based method. Eng. Struct. **25**, 57–67 (2003)
7. Sih, G.C., Tzou, D.Y.: Mechanics of Nonlinear Crack Growth: Effects of Specimen Size and Loading Step, pp. 155–169. Martinus Nijhoff Publications, Boston (1984)
8. Bovsunovsky, A., Bovsunovsky, O.: Crack detection in beams by means of the driving force parameters variation at non-linear resonance vibrations. Key Eng. Mater. **347**, 413–420 (2007). Damage Assessment of Structures VII
9. Andreaus, U., Casini, P., Vestroni, F.: Nonlinear features in the dynamic response of a cracked beam under harmonic forcing. In: Proceedings of the ASME International Design Engineering Technical Conferences and Computers and Information in Engineering Conference—DETC2005, 5th International Conference on Multibody Systems, Nonlinear Dynamics, and Control, vol. 6, pp. 2083–2089 (2005)

10. Doughty, T.A., Leineweber, M.J.: Investigating nonlinear models for health monitoring in vibrating structures. In: ASME International Mechanical Engineering Congress and Exposition, November 2009
11. Wowk, V.: Machinery Vibration Measurement and Analysis. McGraw Hill, New York (1991)
12. Doughty, T.A., Higgins, N.S.: Effect of nonlinear parametric model accuracy in crack prediction and detection. In: SEM Annual Conference & Exposition on Experimental and Applied Mechanics, Indianapolis, 2010
13. Doughty, T.A., Dally, M.R., Bacon, M.R., Etzel, N.G.: Nonlinear model tracking for varying system geometries. In: SEM International Modal Analysis Conference, Garden Grove, 2013
14. Doughty, T.A., Hector, M.J.: Numerical enhancement of nonlinear model tracking for health monitoring. In: SEM Annual Conference & Exposition on Experimental and Applied Mechanics, Orlando, 2015
15. Doughty, T.A., Davies, P., Bajaj, A.K.: A comparison of three techniques using steady-state data to identify nonlinear modal behavior of an externally excited cantilever beam. J. Sound Vib. **249**(4), 785–813 (2002)
16. Crespo da Silva, M.R.M., Glynn, C.C.: Nonlinear flexural-flexural-torsional dynamics of inextensional beams, II. Forced motions. Int. J. Solids Struct. **6**(4), 449–461 (1878)
17. Rao, S.S.: Mechanical Vibrations. Pearson Education, Upper Saddle River (2011)
18. Doughty, T.A.: System identification of modes in nonlinear structures. Ph.D. thesis, Ray W. Herrick Laboratories, School of Mechanical Engineering, Purdue University (2002)

MIX
Papier aus verantwortungsvollen Quellen
Paper from responsible sources
FSC® C105338

If you have any concerns about our products,
you can contact us on
**ProductSafety@springernature.com**

In case Publisher is established outside the EU,
the EU authorized representative is:
**Springer Nature Customer Service Center GmbH**
**Europaplatz 3, 69115 Heidelberg, Germany**

Printed by Libri Plureos GmbH
in Hamburg, Germany